The Proving Grounds

BENJAMIN O'BRIEN

The Proving Grounds

A Catholic on Ladder 17

IGNATIUS PRESS SAN FRANCISCO

Cover photo:
©freepik_firefight_candid-image-photography_70089

Cover design by Enrique J. Aguilar

ISBN 978-1-62164-742-3 (PB)
ISBN 978-1-64229-326-5 (eBook)
Library of Congress Control Number 2024952845
Printed in the United States of America ♾

Before him no creature is hidden, but all are open and laid bare to the eyes of him with whom we have to do.

Hebrews 4:13

In memory of Paul Sanders and Janine Lieu

CONTENTS

I

SMOKE VISIBLE

The staccato clicking of the station printer jolted me into consciousness. Still heavy with sleep, my brain struggled to put thoughts together. Where was I? Oh, yes, still at work.

I must get to the printer, I thought.

The desk lamp illuminated the page as I forced my eyes to focus. I felt a surge of adrenaline as the words leaped off the paper. This would not be a routine call. I keyed the station mic and announced over the speaker system:

"Engine 6, we're going to 281 O'Donnell Street. Cross streets are Industrial and Charles. Smoke visible."

Then silence. I needed that precious calm before the storm, the thirty seconds allotted to the driver of a fire truck in the time it takes his crew to get to their boots. I glanced quickly at the map. Yes, best to take the parkway. What time was it? Just after midnight.

Okay, I thought, *there won't be any traffic.*

Doors were banging now, and feet sounded on the bay floor. I pushed open the door of the watch room and ran across the concrete to the waiting truck. Climbing into the driver's seat, I fired the ignition.

"Smoke visible?" My captain's voice was calm in the seat beside me. I nodded. Behind me, the two crew members were pulling on their boots and bunker coats.

"Know where you're going?" one of them asked. He was a ten-year firefighter with an encyclopedic knowledge of the district.

"I think so," I replied, "just shout out the turns when we're getting close." Truck doors banged shut. The giant garage door was wide open in front of me. I released the air brake, stamped on the gas, and the fire truck roared into the night.

The human brain has an uncanny capacity for taking a step back from a dramatic situation. Looking back, it was almost as if I were watching the events of that night in a movie. It was happening, yet I was not quite part of it. I have an impression of a dark stretch of road in front of me, red and white flashes from the truck's emergency lights striking the sides of buildings as we raced through the sleeping city. I remember feeling oddly serene as bits of radio traffic broke the silence in the cab. A female voice:

"Engine 6 from Dispatch, be advised, we are receiving multiple 911 calls on this. Reports of smoke in the upper floors. Possible victims trapped."

The captain swung his head around to address the men in the seats behind us. "Okay, guys," was all he said. There was a note in his voice I recognized. This was the real thing.

"Dispatch, this is Car 5." The voice of our District Chief was also tinged with a note of urgency. "Let's get some more units on this. I'd like a working fire assignment put in for 281 O'Donnell. We will be in high-rise tactics."

"Copy, Chief, you have Engine 6, Engine 7, Engine and Ladder 5, Engine and Ladder 3, Rescue 2, Car 1 and Safety 1 responding with you."

Where is my hydrant? I thought.

Memories of tough drill sessions forced their way into my consciousness. "Get your crew water or they're dead," I had been told more than once. Water supply was the driver's responsibility, and an all-important one. More memories flashed into my mind. Less than a month previously, we had gone to this same complex on a routine alarm call. I recalled that there was a hydrant to the right of the parking lot. But where to stage the truck? Oh, yes, directly in front of the standpipe, to the left of the main lobby. We swung off the parkway onto O'Donnell Street. My foot came off the accelerator as my eyes scanned for the address.

"Keep going!" someone in the back yelled. Another block farther. Almost there. Suddenly, to the left, loomed the unmistakable bulk of a residential high rise. A heavy column of smoke was drifting into the night sky, glowing dully orange in the city lights.

"Dispatch from Engine 6!" My officer's voice was strained, but I could sense the self-control that was keeping it in check. "We're on location. You can confirm a working fire. We have heavy smoke showing from sides 2 and 3. We will be in fast attack mode. Engine 6 will be O'Donnell Street Command."

I urged my truck across the final stretch of pavement and skidded to a standstill beside the building. Training routines cycled through my mind, mantras that I had repeated to myself over and over again until they had become burned into my memory: "Put it in neutral. Set the brake. Put it in pump gear. Then put it in drive. Watch for two green lights. Listen for the pump to engage." I heard compartment doors slamming. The two firefighters, with the smooth urgency of practiced veterans, were gathering their tools and shouldering hoses. They hurried off after the captain, as he strode toward the waiting building. I jumped up into the open space between the cab and the pump panel. The soft glow of electric lights illuminated the row of levers and gauges. I pulled hard on the handle of the oil primer and clenched my teeth as the noisy grinding smote the air. I could hear the whoosh of water as it was drawn from my on-board tank into the impeller of the pump. Stepping off the truck, my eyes took in the standpipe connections, less than fifty feet away, and the hydrant, standing over a hundred feet away across the parking lot. A Spartan Pumper carries five hundred gallons of water on board, but without a hydrant, even that seemingly large amount can be used up in less than three minutes. Before catching the hydrant, the attack line must be laid. At a high rise, this means a hose must be connected to the building, to supply pipes with water for the interior crews. Quickly, I ran to the back of the truck and began pulling off the 2.5 inch-diameter hose. It was the work of a moment to connect the free end into the brass coupling on the wall and the other end into my pump.

It was then that I became aware of being completely alone. My crew had disappeared, and the sidewalk was eerily vacant. Lazy sparks drifted down on the summer night air. There were no crowds, no screams, no panic. I was like a lonely mannequin performing frantically on an empty stage. A moment later, they started coming. I had not seen the stairwell doors to the left of me, but as they swung open, people began staggering out, at first in twos and threes, and then in a steady stream. Young, old, in various stages of dress, many in pajamas, and a few with smoke-stained faces, a telltale sign that the fire was a deadly one. A middle-aged woman hurried up to me, as I stood poised with a second length of hose in my hands. She pointed up at the building and said with great urgency:

"The fire's on the 14th floor! The 14th!"

"Okay, ma'am. Okay." I nodded emphatically. I could hear sirens in the distance. "We're on it. There's a crew inside." She turned and hurried to join the crowd of onlookers. I hooked up my second line to the standpipe. The sirens reached a fever pitch then shut off. Air brakes hissed, cab doors thumped. I heard radio chatter.

"Command, Engine 7 on location. You can show Engine 5 on scene as well. We'll make our way in and assist Engine 6 with fire attack." The report sounded as calm as if he were discussing the weather.

"I got your hydrant!" Engine 7's driver was at my elbow.

"Thanks!" I said gratefully. The large 5-inch diameter hose slid heavily out of the bed as he sprinted to the hydrant with the free end over his shoulder. Engine 5's driver ran to join him, and together the three of us worked feverishly to move the unwieldy length of rubber. More trucks were arriving. The parking lot was full of firefighters, moving resolutely toward the doors in groups of three and four. With one last click, the supply hose was attached to the intake valve on my truck. I held my arm up and twirled a finger in the air, the signal to send water. Quickly, the firefighter at the hydrant bent over and began spinning the wrench with rapid determination. The hose whipped and snaked as it filled. Regaining the pump panel, I pulled the levers and sent water shooting into the building, up the pipes, and up to the crew battling the fire above. They would need to hook the hundred feet of hose they carried into the nearest standpipe. I knew that it would be black as night up there in the smoke. No one would be able to see his hand in front of his face. They would have to the find the apartment where the fire was lurking, groping their way down the baking, pitch-black hallway. Sightless, they would force the door, search the room, extinguish the fire, and drag out any victims, all before they ran out of air or the heat overwhelmed them.

"All crews on scene, all crews on scene!" It was the District Chief. "Car 5 will assume command. Engine 6, I have you as Fire Attack. Can you give me an update?"

The captain's answer came muffled through his mask: "Yeah, Chief, we're on the 14th. Heavy smoke conditions here. We're still looking for it!"

So they hadn't found the fire yet, and time was ticking. I busied myself with odd jobs around the truck. The first five minutes of a fire are usually chaotic for a pump operator. After water supply is established, however, one can only "hurry up and wait". I bustled about, looking busy but straining all the while to hear my crew's progress. My radio was clasped tight in my hand. At last, an update came.

"Command from Engine 7. We've found the seat of the fire. It's in Apartment 1402. We're just knocking it down now. Engine 5 is coming in to do a primary search." A few moments of silence, then an urgent blast over the radio: "We've found a victim! We're bringing him down!"

I launched myself into the cab and grabbed for the medical equipment. Throwing the strap of the bulky, red first-aid kit over my shoulder, I seized the defibrillator by its green handle and with my one free hand lifted out the oxygen cylinder in its bright orange case. Moments later I was in the lobby, sizing up the scene. A crowd of firefighters, masked and smoke-stained, was performing rapid compressions on the limp body of a middle-aged man. My equipment was seized by eager hands. More hands began slapping the defibrillator pads on the victim's bare chest, the wires tracing back to the waiting machine, ready to shock his heart into life. A clear plastic mask was placed over his face, oxygen blowing into the lifeless nostrils. I noted the pallor of the skin, the dark stripes of soot around his nose and mouth. I wanted to jump in, but the crowd was too thick.

"Give them space!" Someone said.

"Make a hole, guys!" an officer barked. "Medics are trying to bring the stretcher."

For a moment I considered staying, but I knew I would only be in the way. With an effort of the will, I returned to my post at the pump.

It was the next day. I was sitting in a deck chair on the front porch, feeling the August sun warm on my face and watching my daughters playing in the green grass. Everything seemed so serene, so natural. Did the events of last night really happen? I felt for a moment that I must be living two lives. Or was it that there was one life, but two of me? No, it was real. A lingering feeling of accomplishment crept over me. At five years on the job, I had been to many emergencies, but this had been my first time pumping at a high-rise fire. A crusty veteran had thumped me on the back with a grin and said, "thanks for getting us water." The fire had been a moment of proving, and I had stood the test.

It was only later that I learned the man we rescued had been pronounced dead in the hospital. Still later, we listened to a recording of the radio communications from the incident. As I heard again the updates of my crew searching for the fire, questions began to surface inside of me. Why did it take so long? As it turned out, the fire alarm panel in the front entrance showed the fire on the 19th floor, leading them to spend several minutes in fruitless search while it quietly burned on the 14th. But had my actions played a part in the delay? At first I didn't know why I asked myself that question. Then, with a twinge of conscience, a memory emerged, a memory of a woman pointing up at the building, trying to tell me what floor the fire was on. I had stood there with a radio clipped to my belt, a radio I had never thought to use. I had dismissed her, focusing instead on what I had been trained to treat as the most urgent task: water supply.

Today, firefighters operating at a high-rise call perform a 360-degree survey of the building and give a radio report. Interior crews are informed in advance what floor and what side of the building the fire is on, in addition to any visible smoke conditions. Water, we are told, is not the primary concern. Most buildings have wet standpipes, that is, a built-in water supply powered by a strong pump. The role of the fire truck, then, is merely to supplement this pressure. Yet I had been trained from my first day in recruit class to set up a water supply with desperate urgency. Indeed, at a house fire, such urgency is crucial. Crews run out of water in less than three minutes, and there are no standpipes to rely on. That night, however, I had fought a high-rise fire like a house fire. Only in hindsight did I learn that I may have held the key to a vital communication, a communication that could have helped my crew, if I had only taken a moment to listen.

Firefighters learn to live with the consequences of their choices. Often, these choices are made under great pressure. Sometimes, big things depend on seemingly small decisions, even ones that are hidden from others' eyes. I will never know what the outcome would have been if I had given a radio report. Many factors are at play on a fire scene. In all likelihood, the victim was already dead when we pulled up. Still, I cannot help but wonder what the result would have been if I had helped my crew get to him sooner. It is in moments like these that I recall the words of one of my training officers, who at the start of my career gave me this curt piece of advice:

"Don't get attached. The next call is just around the corner."

He was right. One cannot get lost in speculation or bogged down in a cycle of second-guessing. I have found peace in the knowledge that I did my best in a high-pressure situation, acting according to my training and experience. Our performance is never perfect, but we must strive every day to make it better. Indeed, the next call is just around the corner. We learn from our mistakes and move forward.

2

BEGINNINGS

"Push-up position!" Training Officer Miller barked. Four rows of sweating recruits dropped to the pavement. Another day at the fire department training center had begun. Eleven weeks into boot camp, with three to go, I was beginning to wonder if I would make it to graduation.

"Hold!" came the command. A bead of sweat rolled off my nose and splashed on the asphalt. He had been pushing us hard for what seemed like an eternity. On either side of me, young men in buzz cuts, sweat soaking through their T-shirts, were straining to hold themselves stiff as boards. My arms trembled a little. Miller's footsteps beat up and down the row, and his voice beat on our ears.

"There have been reports," he said, "of tasks performed sloppily and in an untimely manner. This is a little exercise in motivation."

He stopped a few yards away, eyeing one of the recruits.

"Malone, get your butt down! You're supposed to be in push-up position. All of you, straighten out!"

I straightened and felt the vibration in my arms increasing.

"Everybody give me fifteen! Go! *One, two, three . . .* "

His voice rapped out the count as twenty-one sweating bodies rose up and down in unison.

"Stop!" he barked. "Into your gear!"

Each man's protective pants, coat, boots, helmet, and

gloves were laid out carefully in front of him. I kicked off my running shoes and stuffed my feet into my boots. Bunker pants were always left with the boots still in them, with the pants rolled down around the ankles, so that a quick pull on the suspenders would dress the firefighter to the waist. Next, I had only to scramble into my coat, don the helmet, and shove my hands into the leather gloves. Woe betide the recruit who was last to get dressed.

"When you're dressed, get into push-up position!" Miller ordered. I finished donning and dropped to the pavement again. I hadn't been the last. The added weight of the heavy suit pushed my body toward the pavement, and I strained to prevent my middle from sagging downward. A burning sensation was starting in my arms. There was a scuffling behind me as a straggler fought with his gear.

"Let's go, Pelowski!" Miller's voice was rising in volume. "Your buddies are waiting for you!"

His footsteps sounded up and down the rows again. Now I was fighting a rising nausea. How long was this drill going to go on?

"When you're in kindergarten!" Miller went on, his voice taking on a menacing tone, "people have lots of patience with you. Here at the Fire Department, we don't have the same kind of *patience* for people who *dawdle* and *take their sweet time*. While you're screwin' around with your coat and can't find your other glove, your buddies are *suffering*. When you don't do your job, your buddies *suffer*. You people have got three more weeks here at the training center, and then boot camp is over. When you get to your first station, they're not going to have patience for *kindergartners* who think they're on a *picnic*. If you're not pulling your weight, you'll be told immediately, and in plain English!"

I was definitely sagging in the middle now.

"Fifteen more, let's go!"

I forced myself into the rhythm and felt my strength waning. The last push-up was a struggle, but I managed to lift myself off the pavement before slumping down to catch a precious second of rest.

"Out of your gear!" came the next command. I forced myself to my feet, peeled the outer layer off and stood at attention, feeling the chilly November wind bite through my wet clothes.

"Moriarty, is that underwear?"

I couldn't see Miller, but his voice behind me sounded incredulous. *Oh, no,* I thought. During firefighting exercises, recruits often wore boxers inside their bunker gear, but today we had been instructed to report to the parade ground in PT gear: shorts, running shoes, and fire-department-blue T-shirts. Apparently, Moriarty had missed the memo. He seemed determined to show Miller that he was not intimidated.

"It sure is!" he sang out cheerfully.

"Yes, Captain!" Miller's infuriated voice rose. "You say, 'Yes, Captain' when you address your training officers! You were told to wear PT gear! All right, more push-ups!"

We hit the pavement again.

The next day, I sat in a straight-backed chair on the second floor of the training center, while two officers eyed me across the table with serious expressions. On the wall behind them was a slogan, stenciled in large black letters: *When Challenged by Adversity, One's True Character Emerges.* Graduation was near, and today was my final performance review. I would learn the truth about whether I had been successful. The two were an oddly matched pair. Captain Bannon was a fiery Irishman with a bright red mustache, the typical "old school" firefighter. Captain Dove was of the newer generation. She wore her hair in a military cut and had numerous tattoos up each arm. She addressed me now in her deep voice:

"When I saw you the first day, I said to myself, 'he doesn't look the part.'"

I wasn't sure if I was supposed to respond to this. I said nothing, waiting for the next remark.

"You surprised the heck out of me, actually," she continued. "You've certainly demonstrated how well you can do this job."

"Thank you," I said, surprised and relieved.

"Yes," Captain Bannon put in gruffly. "There are two kinds of strength, gym strength and farm strength. You have farm strength. Your first time putting up a ladder, I remember thinking, 'he made that look easy.'"

"But you don't look good doing lunges," Dove interposed.

"About your performance," said Bannon. "Very good overall. This is a challenging program, and it tests everybody. You made your mistakes, but we saw you overcome them. You're also very quiet, I notice. We've had trouble with loudmouths in your class, and I appreciate your attitude."

"When you get to a station," added Dove, "the guys will respect that. You have a good work ethic, and you don't make waves. Do what you did here, and you'll be fine."

Three weeks later, in dress uniform, we were marching into the auditorium at City Hall to the drone of bagpipes. We stood at attention and recited the firefighter's oath, pledging to dedicate our lives to serve and protect. Then, each man took a turn walking across the stage, greeting the Fire Chief in a stiff salute, and receiving a diploma. We had graduated from recruits to probationary firefighters. With graduation came my first station assignment. I was to report to Station 12, in the west end. It was a busy station, and I felt a sense of anxiety as well as a thrill of excitement. Life was about to change. The last fourteen weeks had been intense enough. My second daughter had been born the day after I got my call up from the fire department. After a brief recovery, my wife and I had scrambled around, found a house, packed up our things and brought our little family to a new city, a new life, just in time for classes to begin. Five days a week I had learned to force doors, deploy hoses, raise ladders, and search smoked-out buildings. I had waded through stacks of paper on Standard Operating Procedures, fire behavior, and building construction. I was underweight from sweating in bunker gear, doing push-ups, and scrambling to obey irate officers. Finally, it was coming to an end, and I was to begin my life as an urban firefighter.

Probationary firefighters, or "probies" as they were called, had to prove for a year that they were fit for the job. Our contract stated that a probie could be fired at any time, for any reason, at the discretion of the chief. Every three months, I would be moved to a new station, to be evaluated by the captain and crew there. Getting a station like 12 was at the same time a compliment and a challenge. Recruits who performed well in drill school were typically favored with a busy station, but the officers and men in such places tended to be more demanding and less kind to rookies. I knew I would have my work cut out for me to earn their respect.

At ten after six on a snowy December morning, I pulled into the station parking lot and drew a deep breath. The cold wind bit into me as I unloaded my heavy duffle bag and walked with uncertainty toward the double doors. Pushing through them I was met for the first time by the unmistakable smell of a fire station. That smell, so new to me then, later had the power to bring back the keenest memories, some pleasant, some acutely painful: a unique blend of floor cleaner, composite hose, bunker gear detergent and the ever-present hint of smoke. I was in a narrow gear room at the back of the station. Brown coats with reflective yellow stripes hung in a neat row along the wall. Above them, a dozen black helmets lay in quiet formation. On the floor sat a row of leather boots, with pants turned down around the ankles. A brass pole passed through a hole in the ceiling, through which, I guessed, lay the sleeping quarters. Another pair of double doors stood before me, and through the reinforced glass I could just glimpse the fire trucks waiting silently on the bay floor.

"Here we go," I said to myself, and stepped through them.

I checked my riding assignment on the duty board and placed my boots on the floor beside Engine 12. As I was hanging my coat on the door handle, I was introduced to the first of my crew mates. He was a dark-haired, solidly built young man of medium height, and his face wore a grim expression. I climbed up into the cab and sat down beside him.

"Hi, I'm Ben," I said, offering my hand with feigned self-confidence.

"Derek," he grunted, shaking it briefly. I looked around the cab. It was arranged so that equipment could be grabbed in a hurry. A shelf to the left of me housed the medical gear: first-aid kit, oxygen, and defibrillator. A compartment directly in front of me was filled with map books, boxes of medical gloves, bottles of hand-sanitizer, and a bandage kit for burn victims. Flashlights lay in automatic chargers on the center console. Derek was checking his breathing equipment, and I followed suit. Firefighters must bring their own supply of air into fires, and each man's tank, mask, and harness must be checked carefully at the start of each shift. This assembly, known as the "Self-Contained Breathing Apparatus", or SCBA, was carried in a special bracket on each seat. Firefighters could buckle themselves in en route, much like fastening on a parachute harness, and arrive at the call fully dressed and ready for action. I checked mine carefully, turning the knob to the "on" position and watching the tank gauge to make sure I had a full supply. I donned my mask, took a breath to make sure I had airflow, checked the hoses for leaks, then turned everything off again. I rummaged in the compartment in front of me till I had found the logbook. Feeling at my breast pocket, I realized I hadn't brought a pen with me.

"Here," said Derek, passing me his own. "Give it back when you're done."

He began to climb out of the cab, then paused.

"Always carry a pen," he growled, and was gone.

I busied myself with a thorough familiarization of the equipment. The rookie's job was to check the essentials first thing in the morning: All nozzles, defibrillator, oxygen, medical bag, hydrant bag with all the connectors, axe, and of course the Halligan bar, a special prying tool for forcing doors.

As I was proceeding, slamming doors and banging lids to advertise that I was doing my job, an unusual-looking person walked onto the bay floor. At first glance I thought he must be in his teens, but a closer look at his face showed him to be closer to forty. A mop of blond curls partly obscured his eyes, and his short, stick-thin figure looked oddly out of place in a fireman's uniform.

"Hey man!" he hailed me. "You must be the new rookie. We heard you were coming. Welcome aboard."

We shook hands. "I'm Cam," he grinned, "James Cameron Golden actually, but you can call me Cam. You getting all settled?"

"I think so," I replied, instantly at ease.

"All right, well, any questions just feel free to ask."

What a friendly guy, I thought. I couldn't help comparing his reception with the crusty greeting I had gotten previously. It was an early introduction to the spectrum of personalities found in a fire hall. That morning I felt I had seen both extremes.

My first month was a thorough immersion into the life of a firefighter. We didn't fight fires every day, or pull any mangled bodies out of wrecked cars, but I went on my first calls and learned the station routines. We responded to alarms at several high-rises in the district, and as we roared down Park View Avenue with air horns blasting and sirens wailing, I got my first taste of the thrill that accompanies emergency response. I learned to carry a hundred feet of tightly packed hose without letting it slide off my shoulder and to run up stairwells after the captain with a heavy axe and Halligan bar in the other hand. I learned to silence fire alarm panels, reset pull-stations, and what tools to grab at a chimney-fire. I administered oxygen therapy to two unconscious patients and forced a door to reach an elderly woman who had collapsed. I felt I had made a good showing my first month.

One of the things I learned to be familiar with was the station alarm tones. When Engine 12 or Ladder 12 was needed, the station printer would begin automatically spitting out a piece of paper with the call details on it: which trucks were to respond, the address, the nature of the call, and any additional details from the 911 dispatcher. Then, depending on what type of call it was, a loud electronic alarm would sound throughout the station. A long flat tone meant a fire, fire alarm, or car accident; a series of staccato beeps meant a medical call; an alternating high-to-low sequence of notes indicated a carbon monoxide alarm, odor of gas, or other "priority two" emergency that was considered less urgent. No matter what we were in the middle of, we would drop everything and run for the trucks. We had two minutes from the time the alarm went off until the truck should be rolling through the garage door. Even if someone was on the toilet, they would have to cut short and run for it. There was no time to waste.

The pole was the easiest and quickest way to get to the trucks from the second floor. The dormitory was upstairs, and beds had been provided by the city for firefighters on the night shift. Now, with the advent of the 24-hour shift, getting a ration of sleep had become more important than ever, since crews were generally exhausted after a long day of running calls. When the alarm sounded, firefighters would spring out of bed and run for the pole. I learned to be careful whenever this happened. The trap door through which it passed was an occupational hazard. A half-asleep firefighter could easily drop through without gripping the pole properly and end up with a broken ankle. I learned to wrap my legs around it, clench hard, and let gravity take me down. The trick was to let go with my legs before hitting the bottom so that I would land in a crouching position, absorbing the impact. In fact, newer stations were being built with no pole at all to prevent injuries. There had been some close calls at 12. One member was a notorious sleepwalker, and more than once he had been found wandering around the station at night. The crew had designed a special removable cover to go over the hole at night.

In a sense, the calls were the easy part. I was quickly learning that a fire station had a culture unlike any other workplace. One of my training officers had described it as a "wolf pack." It was a good analogy. Privileges were granted based on seniority, and naturally I took the lowest place in the pecking order. Merciless razzing accompanied everything I did. If we were working on a task together, the others would often break early just to see if I would be the last to stop, and drop witty, cutting remarks as they watched me struggle on. While the others were allowed to socialize, exercise in the gym, and even work on personal projects between calls, a probie must keep working virtually every second. When I wasn't being quizzed on my firefighting knowledge or pushed to the limit in training sessions, I would be washing windows, scrubbing floors, counting inventory, and generally keeping as busy as I could. I knew that how I performed at my first station would set my reputation for the remainder of my career. Still, I knew enough to recognize that this was a temporary situation and that the teasing was mainly in good fun. They were testing me, it was true, but if I was reading the signs right, I was also earning their respect.

My other duties included cleaning washrooms, hoisting the flag, keeping a fresh pot of coffee brewing at all times, and, of course, the dishes. Probies had to earn their keep in the kitchen as much as anywhere else, and dishes were the rite of passage. There were unwritten rules associated with this ritual, and I was left to figure them out on my own. A probie could never surrender his territory. He must be prepared to fight, to the death if necessary, to defend it. At mealtime we served ourselves and began eating in order of rank. Hence, I always filled my plate last, and if I wanted to beat anyone to the sink, I had to eat very quickly. As soon as my plate was empty, I would charge to the sink. The others of course would make a game of trying to get there first, and if they did I would have to fight. My best tactic was to step beside them, dig an elbow into their ribs and say, "let me in there." The nicer ones would surrender the sink with a chuckle, but sometimes I would have to resort to stronger methods. A bout of pushing, shoving, and tackling would ensue, with an occasional spurt in the face from the tap if things really got ugly. I tried not to use hot water, though. I had the advantage in height over most of them, but some of the more experienced knew all the tricks, like locking their elbows into the sink and going down into a squat. This made them virtually immovable, unless I got down under their legs and tried to pick them up. In the aftermath, there would be water everywhere, and we would have to mop up the mess. By unspoken rule, the tussling remained strictly good-natured.

The kitchen was the heart of the station. Here firefighters planned their day, discussed successes and failures, and solved the problems of the world. There, I certainly learned a lot about the job, although the advice from my co-workers' personal lives was often at odds with my Catholic faith. I joined in the conversations as I could and would leave the room and start washing windows if topics got too inappropriate. Since the probie wasn't expected to say much anyway, this behavior went largely unnoticed, and I was able to focus my attention mainly on my work.

Another of my jobs was to be the cook's helper. The cook enjoyed a special privilege in the station. He could be a two-year firefighter or a twenty-year veteran, but a good cook was considered an asset to the crew. Some firefighters had managed to stay at the same station for years simply because the captain liked their cooking; others had been transferred over one botched meal. I had been told that the food was "good" on the job, but nothing prepared me for the sumptuous fare I ate each shift. Sausage *penne* was served in a wok the size of a small bathtub, with a rich cream cheese tomato sauce that made the mouth water. Deep-dish lasagnas that could feed an army were downed in one sitting. Steak and caesar salad with mountains of mashed potatoes appeared every so often. But it was the burgers that surprised me the most. I had never seen such massive patties. Each man was allotted about half a pound of ground beef, to which was added sausage, finely chopped jalapeno pepper, diced onions, steak seasoning, and sometimes even coffee grounds. The finished product took about half an hour to barbecue, but when topped with bacon bits, caramelized onions, and melted cheese, it was a creation to outshine anything advertised by a restaurant. I never succeeded in fitting one of these burgers entirely into my mouth; they were simply too enormous. My only frustration was that I had to eat so quickly that I didn't have the leisure to savor such feasts.

I was also responsible for answering the phone. Apparently the penalty for not getting it in two rings was unspeakably dreadful, because I was expected to sprint from whatever corner of the station I happened to be in. In the old days, fire calls came into the station by telephone, and the two-ring rule had been instated. Despite it being the 21st century and the age of sophisticated dispatching, the tradition lived on with all its urgency. One morning, I found myself body-checked in the hall, and from then on it was open war. If anyone beat me to the phone, they were either thrown aside or had to carry on a strangled conversation with the chief while I wrestled them for the receiver. I was not always successful. One trick they played was to call the station from a cell phone while I was doing dishes. When I ran to answer it, one of them took my spot at the sink, and I was forced to fight for it back. The next time it rang, I didn't budge, and the captain came into the kitchen frowning, "how come I had to answer the phone?"

Two months in, I thought I was getting the hang of things. Cam was an invariably upbeat instructor, referring to me affectionately as "Benny", and the rest of the crew seemed to accept that I was now one of them. The district boasted eighty-six high rises, so my knowledge of tall-building protocol continued to develop. There were also several nursing homes in the area, so medical calls were our predominant type of response. I was nervous about treating cardiac arrest, though, and as each shift passed I wondered when I would get my first call of that nature.

My chance came one February morning. I was checking the equipment as usual when the tones went off. Staccato beeps: a medical call. Derek's voice came over the speaker as he read from the print-out:

"Engine 12," he announced, "You're going to 20 Nickel Drive, Apartment J, for VSA."

VSA was short for "vital signs absent", and I jumped for my boots with my heart crashing against my ribs. I dug in the medical compartment and pulled on a pair of latex gloves, donned my safety glasses and loosed the strap of the med bag. The truck lurched out of the station and took off down the street. Moments later, we pulled up in front of a single-story row house. Paint was peeling around the windows, and garbage littered the front lawn. The door to Apartment J was open, and, jumping off the truck with our medical equipment, Cam and I hurried toward it.

"Fire Department," Cam called out as we entered the unit.

"In here!" a woman's voice sobbed. We hurried into the kitchen and found her fighting back tears, her gray hair up in curlers, hugging her faded night gown. She waved an arm wildly toward the back door.

"He's out there!" she wailed, "in the yard. He went out for a minute to shovel snow and just fell down. I can't get him to wake up!"

Through the window, I could see a crumpled figure on the ground. We rushed out; I grabbed his legs, and Cam slipped his arms under the man's shoulders, gripping the limp wrists.

"On three," he directed, "one, two, *three*!"

We straightened. The patient was surprisingly light, and together we carried him into the kitchen. I felt for his carotid pulse and watched the chest for a moment. The skin was still warm under my fingers, but there was nothing stirring below it.

"No respiration, no pulse," I said.

"Okay, start compressions, I'll set up the airway," said Cam calmly, almost cheerfully. I felt for the sternum, placed my palm on it, then with the other hand on top of that one, I began pushing hard. There was a crunch as his ribs cracked under the pressure. I had been warned about this, but it was unnerving nevertheless. I was glad the patient was not awake to feel it. Rapidly, I counted compressions, as Cam inserted a plastic tube into his mouth, attached the oxygen line to the bag-valve-mask, and placed it over the man's face.

"Twenty-eight, twenty-nine, thirty!" I finished. He gave

the bag a calculated squeeze, and the chest lifted. "Keep going, I'll attach the defib pads," he said.

I pushed hard, sweat trickling down my nose. The electronic chirping of the defibrillator filled the room, timing the compressions for me, and Cam worked around me expertly, cutting away the shirt and placing the pads on the patient's bare chest.

"After this cycle we'll switch," he said.

The electronic voice of the defib chimed in:

"*Stop CPR . . . Analyzing . . . No shock advised . . . Start CPR*." The chirping started again, and Cam took over as I moved around to the head. My fingers gripped the mask tightly, keeping it sealed around his face, and remembering my training I kept his chin up and airway open. The captain stood over us, giving radio reports as we worked. The only other sounds were the weeping of the wife in the next room and the endless chiming of the machine. At last, footsteps sounded in the hall, and two paramedics appeared.

"What have we got, guys?" one of them asked, placing his bag on the floor.

"68-year-old male," replied the captain, "VSA on arrival. We've done two cycles, no shock. His wife says he has a history of heart disease and is on medication for cholesterol."

The medics began hooking up their more sophisticated machines, and we paused while they assessed.

"Keep going," one nodded at me, and I dived in for another turn at compressions. After what seemed an eternity, I felt a tap on my shoulder. I looked up and saw the senior medic shaking his head at me with a significant look. I stood up. Cam was spreading a blanket over the inert form, and as he covered the face, I heard the captain talking into the radio.

"Dispatch from Engine 12," he said, "Patient is code black. We've been cleared by paramedics, returning to station."

Code Black was the term used when a patient was deemed deceased beyond resuscitation.

Eternal rest grant unto him, O Lord, I prayed in my mind.

We drove back to the station in silence. I reviewed the call with mixed feelings, glad that I had performed my duties without any major errors, but noticing an accompanying feeling that I couldn't name. I had witnessed my first fatality on duty. As it turned out, I would perform CPR seven more times in my short stint at 12. With the regularity of that type of call came a certain level of comfort with the tasks expected of me. Sadly, most of the victims we treated had been in cardiac arrest for far too long for CPR to be effective. Despite the lack of success in our resuscitation efforts, however, I was gaining confidence in my medical skill set. Feedback from the crew was largely positive, and my earlier fear about VSA calls evaporated. Indeed, life was not all dark incidents, for soon a very colorful individual entered the scene who would bring my focus elsewhere.

3

THE PIRATE

The only way to describe Curtis DeMario was that he was a character. He was short, powerfully built, and sported greasy shoulder-length hair streaked with gray. His raspy voice and tendency to quick action had earned him the nickname "Pirate". Our regular captain was away on holidays, and Curtis was to replace him for a few shifts. My first encounter with him was on a Sunday morning near the end of my three months at 12. Truck checks were done, and I was keeping busy shining the brass pole. The door banged open, and he swaggered in, stopping when he saw me.

"You the rookie?" he rasped. I replied in the affirmative. "Well it's Sunday," he said. "You shouldn't be doing extra work on Sunday."

I grinned a little uncertainly at him and gathered up my polish and rags. He passed into the front office, and I heard him settle into one of the rolling chairs. In a minute, there was the sound of wheels on the floor, and his head appeared in the doorway.

"God made six days for working and one day for rest, Junior. When you work on Sunday, you're offending God."

The head disappeared again, and I heard a little chuckle. I had a feeling that he wasn't religious, despite the sermon. It was a pity, because I would have liked to work with at least one firefighter who shared my beliefs. This was the first time I had a job where I was required to work on

Sundays. My wife and I would take the children to Saturday night Mass, and I knew I would have to keep the spirit of Sunday alive somehow, at the very least in my own heart, even while my duties kept me busy. Oddly enough, the fire department still gave a token nod to the tradition of Sabbath rest, for there was never any training scheduled. Station routine seemed more relaxed, and the other firefighters would disappear shortly after truck checks. During my round of chores, I would spot them reading, watching TV, or working out in the gym.

The Sunday meal was another way they made the day special. The cook would start preparing for it on the Friday shift, laying the foundations for an elaborate spaghetti dinner. He would start by mixing up an enormous batch of ground beef, adding ground sausage, steak seasoning, salt, pepper, basil, and various other spices. I would then be put to work rolling dozens of golf-ball-sized meat balls and arranging them on an oven sheet. Once baked, they were added to a thick tomato sauce, complete with shredded carrots, onions, peppers, mushrooms, celery, and more spices. Served over noodles with a side of garlic bread browned to perfection, it was a festive meal indeed.

Sunday spaghetti had not gone unchallenged in the history of the fire department. A few years before, the city had amalgamated with several smaller municipalities, each with its own fire department. The firefighters of East Sussex had had their own Sunday traditions, including brunch, a very different affair but no less elaborate. East end brunch included copious amounts of eggs Benedict, hash browns, scrambled eggs, sausages, and enormous fruit salads. I had heard stories of how officers from the city had been assigned to these outlying stations and tried unsuccessfully to enforce downtown traditions. One imprudent captain had demanded Sunday spaghetti, which apparently almost caused a mutiny. Flaunting his orders, the East Sussex crew went and bought supplies for brunch as usual, and a particularly bold firefighter then marched into the kitchen, placed a tin of canned spaghetti on the floor in front of the captain, and said: "There's your lunch, Captain."

For the remainder of the morning, Pirate continued to address me as "Junior", inquiring about my background and giving unsolicited advice whenever we crossed paths.

"Don't eat so fast, Junior, you're going to give yourself indigestion."

"When the chief asks you what station you want, give him an answer, don't hem and haw."

"You're not working again, are you Junior? I told you God doesn't like that."

These remarks were usually followed up by a guffaw. Clearly, he was enjoying his role as the crusty mentor with a heart of gold.

Midway through the day, the station alarm went off. It was the long flat tone.

"Engine 12, Seventh and Maple, for an MVC." MVC was short for "motor vehicle collision". Within seconds we were rolling, and I was away on my first call with the Pirate. We drew up on the scene of a two-vehicle crash, and I jumped out of the truck with the med bag. An ambulance was already there, and two young paramedics, one male, one female, were rolling a stretcher toward us. The young man, who seemed to be in charge, addressed me.

"One driver is out of the vehicle, says he's fine. The driver of the one that got rear-ended is still inside, complaining of neck pain. Can you jump in and take spinal?"

I complied quickly and opened the passenger door. A sandy-haired male patient of about thirty was sitting behind the wheel, talking anxiously into his cell phone.

"I gotta go, honey," he finished quickly. "Call you back."

"Hey," I said, "how are you doing? You got neck pain?"

"Yes," he groaned, touching the back of his neck, "right here. That idiot ran into me and made me jerk forward. I think I've got whiplash. It's a brand new car, too!"

"Okay," I said, "keep looking straight in front of you. Don't move your head at all. I'm going to feel down your neck and back. Tell me if anything hurts, even slightly."

I quickly palpated his spine and felt nothing that moved or crunched, no bruising. I placed my hands on either side of his head.

"I'm just going to hold you still, all right, buddy?" I said. "Paramedics are here, and they're going to check you out in the ambulance."

He nodded.

"Don't nod," I ordered, gripping tighter. "Don't move your head."

The paramedics were on the other side of him now, holding a stiff plastic backboard.

"I'm just going to slide this under you, okay?" the senior medic directed. "Then we're going to turn you and get you to lie down."

It was awkward work in the cramped car, but I was able to keep my hands on his head until he was turned. The female paramedic took over, and I hopped out to help with the loading process. With four sets of hands working, we managed to shift the board onto the stretcher, and the paramedics began immobilizing his head with a spinal collar. There was a brief moment when nobody had his hands on the patient. In that second, the backboard, which was not yet strapped onto the stretcher, took a sudden plunge toward the pavement. Hands shot out and stopped it just in time, and the senior medic barked at us with an exasperated, "come *on*, guys!" In a matter of minutes, the patient was safely buckled down and loaded into the ambulance. The young paramedic, however, seemed to have business to settle with us. He walked up to Pirate and frowned.

"Why wasn't the patient collared in the car?" he demanded.

The captain snorted. "My guy was in there holding C-spine! Why don't you go ask one of your people why it wasn't done?" Then, turning to me, he said loudly, "you see why I hate 'em eh?"

I cringed inwardly and glanced apologetically to the medic, but he had turned away angrily and was stalking back to the ambulance.

"Paramedics!" growled Pirate. "They're all the same. And another thing, Junior, he's not your boss. When he tells you to get in the car, look at me first. If I give you the nod, then you go!"

It was my first experience of the turf war between firefighters and paramedics, and I couldn't help feeling ashamed for both professions. Here we were, fellow emergency workers, squabbling like little children. Apparently the altercation was routine business for the captain, however, for he filled the cab with jokes and small talk all the way back to the station.

I was to be treated to yet another display of his quirkiness later that afternoon. At 2:23, we were dispatched to a report of alarms at a nearby retirement home. When we drew up in front of the multi-story brick building, the inmates were evacuating slowly, some with walkers, some in wheelchairs, and nearly all clutching their clothing round them in the cold. Pirate led the way inside, I following with my load of tools and hose. A middle-aged staff member with an ID badge around her neck met us in the lobby.

"One of our residents pulled the alarm," she said apologetically. "There's no fire."

"Let's get them back inside," said the captain, politely

enough. "It's cold out there. Whoever pulled the alarm can wait outside though, haha!"

She hurried off to usher everyone back. Meanwhile, the captain walked over to the lockbox and opened it with his master key. It was empty. He looked around him angrily.

"Who's in charge here?" he barked. "Somebody find me a caretaker!"

A large man with olive skin and a shaved head looked up from behind the main desk. He sized up Pirate for a moment then stood up and turned his back.

"You there!" called the captain. "You in charge? I need keys for the fire alarm panel. You're supposed to have them in your lockbox! Go get me some keys!"

The man ignored him and began sorting papers on the shelf. Pirate swore audibly, and I could see his face turning red. At that moment, the lady intervened.

"I have keys," she said hastily, appearing at his elbow.

"Then you take me there! This guy doesn't want to do his job!"

We crowded into the elevator and descended one floor to the basement electrical room.

"I'm sorry," the woman said, leading us out into the hall, "I'm in trouble, aren't I?"

At once Pirate assumed an air of sympathy and charm.

"No, no, no, sweetheart!" he said soothingly, patting her on the arm, "I'm not mad at *you*!" His face darkened. "It was that idiot upstairs!"

With the alarms reset, we returned to the main floor. The captain advanced menacingly toward the desk where the caretaker still lurked in sullen defiance.

"Okay, buddy, you see the red hat?" He tapped his helmet. "That means I'm in charge here. You listen to me."

The man flashed him a smoldering look, turned, and walked out of the room.

"Of all the . . . !" Pirate fumed, speechless. Then an idea seemed to strike him. He pulled out his radio.

"Dispatch from Engine 12!" he snapped.

"*Go ahead, Engine 12.*"

"I need a District Chief on scene. Building staff are being non-compliant."

"*Copy, Engine 12, we'll dispatch Car 2 to assist you.*"

He stormed outside. We waited on the sidewalk while he raged aloud.

"That son of a . . . ! Who does he think he is? Doesn't want to do his job! He just doesn't care! Just doesn't care! Wait till there's a fire! Then see him come whining for us! What a loser!"

Presently, I heard sirens in the distance. A chief's car crested the ridge and came racing down the street toward us.

"Here comes Bill!" cried Pirate. "He'll set 'im straight!"

Car 2 drew up beside us, and the window rolled down. I saw the chief regarding Pirate with a mixture of concern and amusement.

"Everything all right, Curt?" he asked. "You didn't lose it again, did you?"

"No, no. I was nice. That guy in there just needs a talking-to."

"Well, I thought I'd better come lights and sirens before you did anything violent."

"I wouldn't do anything like that! That's why I called you! Go read him the riot act!"

The chief, a short, white-haired man with steady gray eyes, got out of the car and strolled toward the building, hands in pockets. A minute later, he was back.

"They'll have a key put in today," he said casually. "You guys can return. And Curt," he looked at Pirate, a smile playing at the corners of his mouth, "next time don't scare the old ladies."

4

BACKWATER

After my intensive three-month initiation at No. 12, I was notified that I would be moving to Station 15, on the other side of the city. Cam seemed genuinely disappointed for me, and he shook his head, saying: "It's too bad. 15 is one of the worst stations in the city."

"Worst?" I said, "How so?"

"It's pretty slow there, Benny, pretty slow. And the worst part is, they stick all the problems there. It's the Island of Misfit Toys."

This was an unusually pessimistic speech for Cam, and I felt a sense of foreboding about the next three months. Later, I happened to mention to my lieutenant where I was going. He laughed loudly.

"Bring a good book," he said. "And don't let Lou get under your skin."

"Lou?" I asked.

"Oh, you haven't heard about Lou," he chuckled. "His real name is Luigi Bonnazzio. Let's just say he's a bit of a character. He means well, but there are a few screws loose." He tapped his head meaningfully.

"I see," I said. I wasn't quite sure what to picture. Was he a bully? A prankster? There was nothing for it but to wait and see.

My last shift at 12, the captain and lieutenant sat me down for a debrief.

"We're very happy with what you've done," the lieutenant began. "I don't think we've ever had a rookie who's worked this hard. The chief expects a written performance review, and rest assured I'll give a glowing report."

"Thank you, sir," I replied.

"And don't worry about getting sent to Station 15," put in the captain. "It's not a punishment. They usually put probies in at least one slow station to see how they manage their free time. Just keep doing what you did here, and you'll be fine. Hopefully, after that, you'll get sent downtown."

"15 is a single-engine hall," the lieutenant explained. "Suburbs. You won't get a lot of calls. Your regular captain is acting in a chief's position all summer, and so the station will be run by acting captains who rotate through every few shifts. There won't be much scrutiny, and you'll be left to your own devices a lot of the time. I'm not sure if there are as many windows to wash as here," he finished with a smile. Inwardly, I resolved to keep as busy as possible, regardless of how many windows there were. He looked down at his paper.

"Your performance is rated out of 5," he said. "I wish I could give you 10, but I've given you top marks."

We stood. I shook them each by the hand.

"Good luck," the captain said.

When I walked into Station 15 the next shift, the first person I met was Lou. He was short and swarthy and wore a permanent, lopsided grin.

"Hi, I'm Lou," he exclaimed, holding out his hand. "You're the new rook, huh?"

"Yes," I replied, shaking hands. "Nice to meet you."

"Yah, you too!" he barked. "Make yourself at home. Now, where the hell did I put my mask?"

He began rummaging in the cab, muttering to himself and

making odd little grunts and sighs of annoyance. I found the kitchen and introduced myself to the captain of the day. His name was Frank, a large, friendly man with a loud voice. In the hallway, I met my other co-worker, a young man about my own age who was working an overtime shift.

"Tim," he introduced himself. "Have you ever worked with Lou before?"

"No," I said, "but people keep warning me about him."

"Don't worry, he's quite harmless. Just don't let him corner you when he wants to tell a story. You'll never get away."

I was beginning to get the picture. I was to be in close quarters, not with a bully or a prankster, but with an eccentric who didn't know when to shut up. All things considered, that wasn't such a bad lot. Not as bad as I had feared, at any rate.

In the early afternoon, the tones went off. "Residential alarms! 886 Cedarwoods," came the announcement. I ran for my boots and saw Lou scrambling into the driver's seat.

"I don't know where that is!" He was yelling, obviously flustered.

"Don't worry, Lou, we'll find it," I heard the captain say in a tone of reassurance. Right away I felt a lessening of my own blood pressure. I could tell that this was a good officer. Lou seemed to settle down, too, and Frank directed him calmly through the streets with brief glances at the map. At one point, the truck tipped and came back down with a crash as we rolled a curb. Tim leaned over and said in a low voice: "Twenty years on the job, and he still can't drive a fire truck."

The call turned out to be a false alarm, and we returned to station. It was free time, and Lou and Tim made their way to the lounge. I familiarized myself with the truck for the second time, then wandered in to join them. Tim was telling Lou all about his holidays.

"Just got off the plane last night," he was saying. "I'm lucky I was awake when the chief called to offer me the shift. Overtime is a great thing!"

Without warning, Lou burst out: "Don't talk to me about overtime! What a bunch of crooks up there in headquarters! I've been screwed over so many times. They make you think they're on your side, then they stab you in the back! That's what I said, stab you in the back! I told the chief, I said to him, right to his face I said, Chief, you're not innocent until proven guilty. No sir, you're guilty until proven innocent. That's what I said! And now I'm in his bad books, but I don't care!"

He finished with a croak of a laugh. There was an awkward silence, and Tim and I looked at each other, not sure what to make of his speech.

"Well," said Tim finally, "I think I'll go work out now." He stood up and left the room, and I followed. I changed into my shorts and running shoes, and joined him in the gym.

"Did that just happen?" I said, incredulously. He shook his head with an amused smile.

"I think he was actually mad at me. The problem is, I have no idea what about!"

The next shift, I met the other crew member I would be spending my time with. Jacques stood six feet, four inches tall, and where Lou wore a perpetual grin, he wore a perpetual scowl. He had been disciplined frequently for insubordination, I had heard, and had a tendency to neglect essential duties like checking the truck. I sensed quickly that I would be working with two eccentrics instead of one. I could only imagine how having the two of them in close quarters would play out. Our first meeting was friendly enough, however.

"We're not downtown here," Jacques advised. "You don't need to jump up for dishes or be cleaning every second of the day. If there's something that needs to be done, we'll come and get ya."

He was as good as his word. After supper, I stood and headed for the sink as was my habit before the others had finished eating.

"Sit down!" he ordered. "I told you we're not downtown. Save that for Station 3!"

In the days that followed, he established his territory in the kitchen. He was the self-proclaimed chef, and the first time I wandered in to offer a hand, he chased me out with a gruff but good-natured, "get out of my kitchen!" Lou and he seemed to avoid each other instinctively. Jacques' domain was the "watch", the station control room that housed the alerting system, phones, and radio. He could usually be found there with the TV on and his feet on the desk. Lou, meanwhile, liked to occupy the picnic table in the station yard, chain-smoking and thumbing his cell phone. I decided I preferred Lou's company, and occasionally I would go outside and join him. He always had something to say, and one day it was a benevolent lecture on how to be a good fireman.

"I remember being a rook," he said, leaning in confidentially, cigarette hanging out of the corner of his mouth. He always said "rook" instead of rookie. "Man, I remember the pranks they used to play on me then. Course I've done lots of things to rooks myself. We used to douse them, *electrocute* them, string them up in the hose tower, tie them to their beds!" He paused to cackle. "I loved it. I loved it when

I was a rook, too. I always played along. One piece of advice: You're a rook? You suck up. You suck up hard." He wagged a finger. "What do they tell you now? Keep your mouth shut for a year? No, keep it shut for five years! Then tell them all to go to hell. That's what I do!"

I was doing my best to follow his speech, nodding occasionally. I reached down absently and picked a long blade of grass from a clump that was growing beside the picnic table. I put it in my mouth and began chewing. He stopped mid-sentence and grinned widely.

"Country boy?" he barked.

"Yes," I replied. He gave a delighted chortle.

"Ah, that's great. Straw in the mouth and all that, I love it!"

Later, at supper, I was eating quickly so as to be first at the sink. Lou gave the same wide grin.

"You from a big family?" he asked. I nodded.

"Ha! Thought so! Gotta eat fast or it'll be all gone! Every big farm family I've known, it was the same thing . . . Hey!" he nudged Jacques. "He was chewing a straw earlier. Country boy!"

Jacques grunted and kept eating. Lou chuckled then rasped out a smoker's cough.

"Country boy, he! he!"

He took his first bite of pizza. Suddenly he gave a grunt of surprise. He threw down the piece and sat glaring at it.

"Is this whole wheat crust?" He spat out, leveling an aggressive stare at Jacques.

"Yeah," Jacques shrugged. "What's your problem?"

Lou lurched to his feet and jabbed a finger across the table at him.

"You know I hate whole wheat crust!" he yelled.

Jacques stopped chewing and aimed a level gaze back at him.

"If you don't like it, you can cook next time," he said coolly.

"Oh yeah?" Lou yelped, "well how about this: How about I'm not eating your meals ever again!"

"Go ahead," said Jacques evenly. Lou plopped back into his seat, visibly stewing. I glanced at the captain, who was working away placidly at his own meal. Apparently this was not an uncommon scene. I settled a little lower in my chair, trying to seem invisible.

We caught a fire the next shift. Lou gunned the truck through the streets in a fever pitch of excitement, while I hastily buckled my coat and slipped the straps of my breathing apparatus over my shoulders. We rounded the bend onto Montgomery Avenue and saw a two-story detached home with smoke pouring from the upper windows. Engine and Ladder 14 were parked in front, and as I jumped out of the cab, I could hear the engine crew attacking the door with a sledge hammer. Frank finished his radio report, then climbed out to join us.

"Help them get water!" he called. "Then meet me inside!"

I ran past Engine 14 to where the yellow hydrant stood on the sidewalk. Jacques helped the driver pull off the supply line, while I popped the caps off and threaded the connectors. It took only a moment to hook up the hose, and when the driver signaled to me I spun the wrench and saw the line stiffen as it filled.

"Come on!" Jacques was yelling. "Captain's inside already!"

I gathered up my tools and hurried after him. The captain

was in the foyer, donning his mask. I could hear smashing above me, and a hose-line was stretched up the stairs.

"Get your masks on, guys!" he ordered. "We're going up to do the primary search."

I strapped on my facepiece and turned on the air. With my hand on Jacques' shoulder, I followed him up the stairs and into the smoke layer. It was pitch black. My free hand groped for the button of my flashlight, but even the powerful beam barely pierced the gloom.

"Go right!" I heard Frank call.

I felt along the wall till I came to a doorway.

"Here's a room!" I called.

"Search it!" came the order. I dropped to my hands and knees and moved forward, one hand on the wall, while the other swept the floor carefully with the axe handle. I felt a bed, a closet choked with fallen clothes, and suddenly found myself in a bathroom. Still, moving right, I found the door and emerged back in the bedroom with the smoke beginning to clear. I could see Jacques at the door, shining his light in at me.

"All clear!" I said. He waved me over and led the way to the next room. The attack crew had done their work, and the fire was out. I could just make out the remains of a charred bed, almost hidden under a mound of fallen drywall and burnt strapping. The window was gone, and parts of the wall around where it had stood were missing, too. The fire must have vented out the window, consuming part of the structure with it.

"No one was home," said Jacques. "Good thing, too."

We helped Engine 14 drain and roll their hose, then we climbed back on board our own truck. As we drove away from the scene, Frank turned and looked back at me.

"Good job today!" he said.

Back at the hall, we hosed off the axes, changed our air bottles, and made sure everything on the truck was response-ready again. Dusk was falling, and as I made my way into the living quarters, I heard an unusual sound coming over the speaker. It sounded like an orchestra, and as the volume increased, I could make out a stirring march of some kind. I poked my head into the watch and saw Jacques grinning at me.

"That's the Soviet national anthem," he said. He was holding his cell phone into the mic and playing a recording. "I just thought I'd remind you to take down the flag."

If I ever forgot after that, the Russian anthem would blare over the speakers, and I would hurry to the flag pole. I did wonder whether it was appropriate to use a communist theme song to signal my patriotic duty, but none of the captains seemed to mind.

As the weeks wore on, I became accustomed to my co-workers' colorful ways. I found their quirkiness didn't bother me; in fact, it kept life at a slow station entertaining. And, if I was perfectly honest, the laid back atmosphere of the place was a welcome respite after the unrelenting whirlwind of drill school and Station 12. At last, however, the time came for me to leave. I was going to Station 3, the busiest fire hall in the city. It was nicknamed "The Big House" and was considered a challenging environment in which to prove oneself. Frank gave me my review, as well as a few words of warning before going downtown.

"You certainly managed your free time well," he said. "And the guys enjoyed having you. Just be mentally prepared that it's a very different world at the Big House. It's a weird game they play down there. They'll make you jump to their tune to see what you're made of."

Jacques and Lou shook my hand and sent me on my way

with a few teasing remarks. As I walked out the door for the last time, I paused to look back at the old brick building. I didn't know if it ought to be called the worst station in the city. It was slow, that was true, and full of eccentrics, but I knew I would remember my time there with fondness. It had been a strange but not unpleasant interlude before I was to hit the ground running at Station 3.

5

THE BIG HOUSE

A river of traffic ran along the street beside me. Crowds of pedestrians swarmed the crosswalks, competing for space with cyclists and buses. The old fire station overlooked this bustling scene, faded brass lettering paying tribute to the years it had served the downtown. "Engine Co. 3—Hook & Ladder 3" it read. A group of tourists stopped to snap pictures, as I clipped the flag to its halyard and sent it fluttering up the pole. Another day at the Big House had begun.

I made my way to the janitor's closet and collected my cleaning supplies. Washrooms were to be shined at exactly 9 o'clock. After making my rounds, I checked the coffee pot in the kitchen, wiped the counters, and headed to the bay floor for truck checks. The others were still in the lounge, so I produced a rag and began wiping mirrors and windshields. I looked up as Neil, my driver, approached.

"Do you ever stop?" he asked, shaking his head at me. "Take it easy man, you're killing it here."

I grinned. It was a good sign. We checked our inventory, started the saws, and Neil pulled the massive ladder truck out on the tarmac. He motioned for me to climb up onto the turntable, and I took the controls, extending the 100-foot ladder and putting it through a full rotation. After truck checks, I consulted my list. The captain and lieutenant had sat me down on my first day and laid out the expectations for my stay.

"We give you all the worst jobs," the lieutenant had said bluntly. "This is where you build your reputation. The last probie worked extremely hard, so you've got big shoes to fill. You'll have no spare time. Keep busy, and remember, you've got two ears and one mouth, use them proportionately."

He then handed me a paper, which I was to consult every shift as my road map to success:

Station 3 Rules

—Report for duty in uniform. No workout gear.

—Check apparatus with crew at the start of every shift.

—Learn equipment location on every rig, as you will be the one asked to retrieve it at an emergency scene.

—Check medical equipment.

—Raise and lower flag.

—Take watch duty between 1200 and 1400hrs.

—Answer phones within two rings.

—Greet civilians and conduct tours.

—The spare engine is your responsibility. Perform brake checks and inventory.

—Ensure bathrooms, kitchen, and lounge area are kept in a state of cleanliness.

These rules were straightforward enough. Then came the section that I found myself rereading often:

—Generally speaking, a wise rookie listens much and speaks little. A know-it-all attitude is not good.

—Be the first to volunteer for a job and the last to finish working.

—In the fire service, reputations take a long time to build and a short time to tarnish. Be aware of this, and conduct yourself accordingly.

—Show a willingness to work. People will go out of their way to help you succeed.

—Questioning logistics, whether in the station or on the fire ground, will not be well received. Be prepared to earn your stripes and do as you are told. Asking questions is encouraged, questioning authority is not.

—You are now interviewing for a station.

I folded the worn paper and put it in my pocket. At that moment, the tones went off.

"Ladder 3. Ladder only. 3375 Bryant. Cross streets are Mayfield and Fallingview. In Station 1's district. Smoke visible."

I ran to the pole, wrapped my legs around it, and slid to the floor below. Doors banged as firemen ran to the truck. I jumped into my boots, pulled up the pants, and slipped the suspenders over my shoulders.

"Suckers!" one of the engine crew yelled. The call was out of district, so Engine 3 was not required, only our ladder truck.

"Yeah, you're just jealous!" Neil shot back. "Have fun running medical calls!"

He fired the motor. I settled into the bucket seat, threw on my coat, and felt for the straps of my SCBA.

"Take the highway," I heard Lieutenant Snedden say, implacable as always.

We accelerated out the doors and veered left onto the street, then lurched to a halt behind a wall of stalled traffic. Neil swore and yanked on the air horn. I saw cars squeeze over to the curb in an attempt to make room. We began to creep forward, moving between the parted rows with inches to spare. At last we were through, and Neil was back on the throttle. We rocketed down Stevens Boulevard, through several sets of green lights, then swung onto the on-ramp. The first radio reports were coming in.

"Engine 1 on scene with Ladder 1. We have a single-story automotive dealership, smoke showing from Side 1. You can put in a working fire. Engine 1 will be Bryant Command."

"Copy, Engine 1. Be advised, you have Engine 2, Ladder 3, and Car 1 already responding. We will add a working fire assignment."

"Copy, Dispatch. Engine 2, can you come down Mayfield and get the hydrant on the corner?"

"Yeah, we got your water, Engine 1."

"All units, this is Car 1. I'm assuming command. Ladder 3, you will take the front of the building and establish roof sector."

Snedden turned to give me my orders.

"You and I will be getting up on the roof," he said. "Grab an extension ladder and a chainsaw."

"Yes, sir."

Neil slowed down as we came up to the off-ramp. A few turns, and we were on the fire street. Firefighters were rushing a supply line to the hydrant, and a plume of bluish smoke was issuing from the eaves of a commercial building. I noticed something odd about the place. Where the street passed the building, it turned into an overpass, so that the roof of the business was level with the sidewalk. Snedden saw it, too.

"Forget the ladder," he ordered. "Chainsaw and axe. We'll step right onto the roof."

I hurried to the compartment and hefted out the saw, then seized the fire axe from its bracket and ran to join him. He was standing on the sidewalk calmly sizing up the scene. I was about to step forward when his arm shot out and stopped me.

"Take care!" he said sharply. I looked down. Between the roof and the overpass was a gap of about two feet. I could see concrete at least ten feet below us.

"That's deadly," he said. Then, pulling out his radio, he announced: "All crews, be advised, there's a space between the building and the street. If anyone is working on the Bryant side, be careful!"

He turned to me and motioned to a section of roof that butted up against the roadway.

"Go up that way," he said. I skirted the hazard and scrambled quickly onto the shingles, tools in hand. The roof had only a slight grade, and I was able to stand up quite easily.

"Make a hole," he directed. "Let this smoke out. And we're above the fire, so watch it."

I started the saw and cut a line down the roof, running parallel to the rafters, then turned the corner and began cutting perpendicular. Smoke began to trickle out the cracks. Two more cuts, and a square hole appeared. I worked away with the axe to clear debris, as heavier smoke gushed out. Bodies swarmed around us. The crew from Ladder 1 had arrived with a hose line.

"Here, grab the nozzle," one of them said to me. "I'm gonna open that ceiling."

I put down the saw and hung onto the hose, while he jabbed through the hole with a pike pole. I watched with interest as the smoke intensified. He was breaking the barrier to the fire room below, and we were getting the full brunt of the escaping heat. Snedden grabbed him by the shoulder.

"Careful, there's a crew inside!" The man nodded and

stepped back from the opening. Next moment, the lieutenant was on the radio, giving an update.

"Command from roof sector. We have a vent opening over the fire."

"Copy, roof. Fire attack, did you get that? You have ventilation. You're a go for hitting the fire."

"Copy, Chief. We've just gained access to the room. We'll flow some water."

The smoke changed color as billows of steam rose and mixed with the darker smoke. They had a hose stream on it now.

"Make that hole bigger," Snedden ordered. The ladder crewman put down his pike pole and took up the saw. The motor revved and sparks flew as he cut through roofing nails and gusset plates. All at once, there was a commotion. A panicked voice was coming over the waves:

"Firefighter down! Firefighter down!"

I looked around wildly. There on the sidewalk were two firemen looking over the edge into the narrow gap. One of them was yelling into his radio.

"We have an officer who has fallen off the roof! We need a ladder on Side 1! Now!"

Snedden was sliding down the roof toward the accident. I thumped the firefighter from Ladder 1 on the shoulder.

"Someone fell!" I bawled into his ear over the noise of the saw. He stopped, looked at the crowd that had gathered at the parapet, the men rushing up with a ladder, then back at the hole he was cutting, and shrugged.

"Keep working," he said. "RIT's on it." RIT, or Rapid Intervention Team, was a special engine company set aside at every fire to rescue trapped or injured firefighters. Everything in me wanted to drop what I was doing and join them, but I realized that he was right. Adding myself to the congested scene would only be counterproductive. There was still a fire to fight. I was a little amazed at his indifference though. I gripped the nozzle tighter and watched the drama unfold below me. They had the ladder over the side now, and two firefighters were climbing down. I heard them give an update:

"Command from RIT. We've reached the downed firefighter. He seems to be OK. He's complaining of a sore back, but I think we can get him up the ladder."

In a moment a black helmet appeared, followed by a red one. I recognized it as the captain from Engine 10. He climbed slowly onto the sidewalk and stood uncertainly, while a crowd formed around him.

"Hey!" It was my friend, the chainsaw man. "Wake up and spray that down!"

I opened the nozzle and directed a fog stream into the void space. He poked and prodded with the axe, turning over insulation and ripping out blackened bits of wood. We were a good twenty minutes dousing hot spots, and Snedden had rejoined us by the time the last wisps of smoke had dispersed.

"Is the captain okay?" I asked.

"Yeah, I think he'll be all right. He hurt his back in the fall, and he thinks he messed up his knee. He's going to the hospital to get checked out. He's lucky. That was no joke."

A few hours later, we were back at the station. Our gear was cleaned and stowed, and I was out in front washing windows. As I worked, I thought I heard muffled laughter coming from somewhere above me. I looked up but saw nothing. All of a sudden, a torrent of water fell on me.

Soaked to the skin, I looked up to see Jason and Brad, the station clowns, leaning over the edge with an empty bucket.

"Is that the best you can do?" I yelled. It was a mistake, for immediately a second bucket-full hit me square in the face. I ran inside, looking for a pail or a hose, anything to get back at them, but they got to me first. Breathless with laughter, they slapped me on the back, and I couldn't help but laugh as well at the sight of their delighted faces. Suddenly Brad looked serious.

"You didn't have your phone in your pocket, did you?" he asked.

"I did," I replied flatly then paused, waiting for his reaction. His eyes widened. I could see the prospect of dropping a hundred dollars on a new phone swirling before his vision. I reached slowly into my pocket and drew out the phone, holding it up for them to see. It was safely sealed in a ziplock bag. They burst into fresh laughter.

"I knew you jokers were going to try it sooner or later," I said. "Luckily I came prepared."

It wasn't all lighthearted at Station 3, however. The trouble had started one evening as I was doing dishes. The others were clustered around a cell phone, and I could tell from their conversation that Brad was showing them some inappropriate pictures. Suddenly, he looked up.

"Get in here, Benny! Be part of the team!"

I froze, dishwater dripping from my hands.

"No thanks," I said.

There was a chorus of surprise from the guys.

"Come on!" said Brad. "Don't be shy."

"I better not," I said. "My wife wouldn't like it." It was not what I had meant to say. My horror of pornography went far deeper than simply "being good" to please my wife. It was rooted in my faith and my desire to treat all women with respect, including her. But the words had come out that way, and I was suddenly tongue-tied, conscious of the hostility that had descended on the room. Snedden snorted.

"Oh, shh! Shh!" he mocked. "That's weird." They returned to the phone, their backs to me. I finished the dishes in silence and left the room to find other chores. I could hear their laughs echoing in the hallway.

The next day, Snedden called over the station speakers: "All hands, lecture room."

I made my way to the back of the station, where a narrow room had been set aside for power point presentations. A projector and a large screen had been set up, and we often trained in there on rainy days, going over tactics. When everyone was seated, Snedden plugged his laptop into the projector. An obscene image popped up on the screen. Instantly, my eyes flashed to the floor, and I kept them fixed on a crack between two tiles.

"This is our training for today, guys," said Snedden, in a mockingly serious tone.

"Ben, you better keep your eyes on the floor," hooted Jason. There was a chorus of laughter.

My fingers stole to my wedding ring, and I twirled it absently, still staring at the floor. I heard Brad chuckle.

"Hang onto that ring," he said.

"All right, moving on," said Snedden, switching off the image. "Let's do some real training."

I stared into space, unable to follow the lecture. Self-condemnation ate into my gut. Why hadn't I gotten angry, shown them that I didn't like it, or at least stood up and left the room? The fear of making waves had held me back, and I was deeply ashamed. Later on, I would realize that the incident could have cost him his job. Recruits were not well-educated about workplace law, and I had no idea that I would have been within my rights to press charges. Instead, I sat through the remainder of the training and went about the rest of my shift feeling relieved that the incident was over.

It didn't end there. The watch desk at Station 3 was three sided, with windows commanding a 180-degree view of the street, and my co-workers were in the habit of "checking out" all the women who walked by. Since the station was near a university, there were a large number of young women on the street every day, going to and from classes. The windows were tinted, which allowed the firefighters to look out without being seen, and someone had even gone so far as to bring in a pair of binoculars. The fact that I didn't join in hadn't gone unnoticed, either.

"Best watch desk in the city," Snedden announced one day. "Of course it's totally wasted on you."

Every few days, the captain was in the habit of switching me back and forth between the engine and the ladder. This gave me the advantage of experiencing a wide range of calls and also afforded me an occasional break from Snedden. Station 3 could run up to twenty-five calls in a shift between the two trucks. The ladder tended to go to alarms or smoke visible calls out of district, while the engine dealt with medicals and MVCs. Both trucks would go if it was smoke visible or alarms in district. Between non-stop responses, cleaning, studying, and training, I was usually in a zombie-like state by the time I got home. My wife and I had worked out a system. She would meet me at the door after her own shift with the children, and we would take a moment to silently scan each other's faces to see who was the most exhausted.

"How many?" she would ask, and I would hold up a number of fingers to show how many calls I had run during the night.

"How many?" I would ask in turn, and she would hold up her fingers to show how many times the babies had woken up. Whoever had the most fingers got to nap.

One night, while I was assigned to the engine, the tones went off for a medical. The address was only a block away, and after a short response, we stopped beside a park. A middle-aged man was curled up on the grass, apparently unconscious. A police car drew up at the same time, and a young female officer got out. I gathered up the medical gear and followed the others as they approached the patient unhurriedly. Brad was the first to reach him, and he gave the huddled form a nudge with his toe.

"Hey, buddy!" he called out. "What's happening?"

The form stirred, and the man sat up blinking.

"Whaddaya want?" he demanded, giving his stubbly chin a scratch. His clothing was worn, and a shopping cart with a few meager possessions stood nearby.

"You all right?" asked Brad, "We got called out cause someone thought you were in trouble."

His eyes narrowed as he looked around at us, then he broke into a toothless grin.

"No trouble!" he cackled. "No trouble. Just sleepin'."

The officer, who had come up behind us, now decided it was time to take charge. She stepped forward, thumbs tucked into her bullet-proof vest, with an air of authority.

"You can't sleep here, sir," she said. "It's public property."

The patient gazed at her solemnly for a moment, then rose ponderously to his feet. He jutted his chin out at a defiant angle.

"I ain't doin' nothin' wrong!" he said. "You got no reason to arrest me!"

He swung around and gave us a big wink. We were in on his joke. The lady persisted with great patience.

"I'm not here to arrest you," she said, "but you can't stay. If you need a place for the night, I can drive you somewhere," she offered. "There's the Good Shepherd Institute or the House of Hope, you can take your pick."

"I ain't goin' in no cop car!" he declared, shuffling over to his cart. "I'll go myself!"

"Okay, that's fine, you can walk," she said.

He looked all around him with an air of injured dignity, then announced with great solemnity:

"I ain't walkin'. . . I'm *flyin'*!"

Then he marched away up the street without looking back.

The next call connected with homeless people wasn't so humorous. It was a sunny Wednesday morning, and we were dispatched to a report of smoke visible in the trees on an abandoned lot. When we pulled up, there was no smoke to be seen, and we beat about the bushes, searching for a possible cause. Snedden, Neil, and I were just about to turn back, when a bearded man suddenly stepped out of the shrubbery.

"Hello," he greeted uncertainly.

"Hi," the lieutenant replied curtly. "We were called for report of smoke. You know anything about that?"

"We were just cooking our breakfast," he said shyly. "We had a little fire, but it's out now."

"Can I see?" asked Snedden.

The man led the way into the bushes, where a makeshift campsite had been set up. There was a cardboard shelter, some blankets, and a shopping basket full of miscellaneous items. A ring of stones surrounded the charred embers of a campfire. A woman squatted near the stones, stirring something in a blackened coffee can. She stood hurriedly as we arrived and averted her eyes. Snedden regarded the fire pit briefly, then spoke with an air of reassurance.

"It's all right," he said. "You're not in trouble. I don't see any danger here. You folks have a good day."

He turned to leave. Neil, meanwhile, had poked his head curiously behind the shelter and now returned with a somber expression. When we were out of earshot he said:

"Three kids sleeping in there."

I felt a stab of pity. What could I do to help ease this epidemic of homelessness that plagued the downtown? When I learned a few weeks later that my parish was organizing a group to volunteer at the soup kitchen in that same neighborhood, I knew it was my chance. I signed my name to the list and spent an off-duty morning slicing buns in the back kitchen of the shelter. It was a small contribution, but perhaps, I thought, if everyone did one small thing, we could slowly make a difference. "You will always have the poor with you," Jesus had said.

~

With every spare minute at the station and much of my off-duty time occupied, the three months had flown by with astonishing speed. I began to look forward to the day when I would be given my permanent assignment. One day in late September, the phone rang, and I recognized the voice of the Division Chief on the other end.

"Just the person I was looking for," he said. "Where do you live, east end or west end?"

"East end," I replied.

"Hmm," I heard him hesitate. There was a short silence, then: "I'm going to put you at 17. You'll start there the week after your pump course."

"Thank you, sir."

"No problem, buddy. I try to keep people close to home. Enjoy."

He hung up. I turned and saw that Brad and Neil had been watching me curiously.

"Well?" Neil asked.

"Well," I said, "I'm going to 17."

"Bummer," he replied. "That's a slow station. I know someone who used to work there, and he said they went three shifts one time without a call."

I digested this.

"What about the crew?" I asked.

"It's a decent crew," he replied. "And Captain Peters is a nice guy. You won't mind it there, just hopefully they'll send you to a busier station before too long."

The next week, I was assigned to the training center for my pump operator's course. This week-long session marked the end of probationary year. If successful, I would begin my life as a Firefighter Fourth Class and assume the responsibility of driving the truck at my new station. Some cities assigned a permanent driver, but here there was a rotating schedule, each man taking a turn for a month. New drivers must be ready, understanding in depth how to operate a fire pump and safely drive the heavy vehicles. It would be an intense week, but I was looking forward to catching up with my classmates. The twenty-one of us had been dispersed to various stations across the four shifts, and many of them I hadn't seen since graduation. I was eager to hear their stories about their first year in station. Working straight days that week meant that I would miss my last twenty-four-hour shift at No. 3, but I would have time on Thursday evening, after the course was over, to stop by and say my goodbyes. Bringing in a transfer cake on the last day of work was a time-honored tradition. Since the phone call had come on my last shift, however, I would have to drop it off during my Thursday visit. After some thought, I settled on a large ice cream cake. It should be ample to feed seven guys, even with enormous firefighter appetites. Accordingly, at twenty after six on Thursday evening, I pulled into the station parking lot. My wife and two daughters were in the pickup with me, and I had the cake on the floor, wrapped carefully in a cold pack and towel.

"I'll just be a minute," I said.

"Oh, can't we come in?" my wife asked. "I'm sure the girls would like to see where you work."

I hesitated a moment, thinking of my co-workers. How would they treat a rookie's family?

"All right," I said finally. After all, they were trained to greet members of the public respectfully. The bay doors were open, and as we walked in, the two little girls stared wide-eyed at the enormous trucks.

"This is Daddy's fire station," I said. "Would you like me to lift you up so you can sit in the truck?"

Beth, the oldest of two, nodded vigorously, her eyes still wide. I hoisted her up into the driver's seat of Ladder 3.

"What do you think?" I asked.

"I want down," she said, reaching out her arms for me. I swung her down, and she toddled to her mother and clutched her leg.

"They're big, aren't they?" I said. "Don't worry, maybe next time you'll want to sit up there longer."

I reached for my younger daughter, Kathleen, who was

being held by my wife, but she hid her face on her mother's shoulder and held on tighter. I chuckled.

"Don't worry," I reassured her. "You don't have to . . . It's often this way," I added, smiling at my wife. "We get a lot of kids stopping by the station. The first time they're shy; the next time we can't pry them off the trucks."

"We'll just have to visit you more often," she replied.

I led the way to the living quarters and pushed open the double doors into the main hallway. The captain came out of the watch desk and stopped with a big smile on his face.

"Hello, Mrs. O'Brien!" he greeted. "Hello girls! I'm Mark."

Hmm, first name basis, I thought. He had never let me address him as anything else but "Captain."

"Hello," my wife answered, extending a hand graciously. "I'm Kate. These are my daughters, Beth and Kathleen."

The other firefighters were gathering around now, introducing themselves. Even Snedden stepped forward to shake hands.

"How was your course?" he asked me.

"I passed," I replied.

"Congratulations," said Neil. "Not a probie any more."

"I see you brought your cake," said the captain. "Why don't you all come along to the kitchen."

I stowed the cake in the freezer, while Kate and the girls settled themselves at the long table. Chairs scraped as the others sat down nearby. There was some small talk, and Neil handed out cookies to the children. At length the captain turned to me.

"We never got to do your evaluation last shift. Do you have a minute? The guys will look after your family."

I looked at my wife. She nodded.

"We'll behave," Brad grinned.

I stood up and followed the captain and Snedden to the watch desk. We took our seats in the rolling chairs, and the lieutenant produced a piece of paper.

"We do this with every probie," he began. "This is your performance review. We rate you on a scale of 1 to 5 for each category. 3 is average, anything below that needs work. Above that is out of the ordinary." He paused, scanning the page. I nodded, remembering the process from my previous two stations.

"Let's just say your performance was excellent. No, let me rephrase that. It was very good. You worked extremely hard. In fact, you made the last guy look downright lazy. As far as fitting in . . . " he didn't finish the thought, but I guessed what he was thinking. Our conflict of morals had prevented me from sharing in their camaraderie. Before I could say anything, he went on.

"Okay, let's see . . . punctuality. You were always on time, so I gave you a three. We don't give more for that one." He ran his finger down to the next item. "Uniform. Three as well. It's not like you could wear two uniforms and get a five . . . next one, 'Follows Directions'. I gave you a four for that. You've done everything we asked you around the station, and you did well at the Bryant fire . . . Let's see . . . 'Uses time well'. Oh, yes, I gave you five out of five."

The captain nodded. "We're very happy with your work. Obviously, we watched you a lot. You kept busy, you listened, you learned, you asked questions. Personally, I'd be happy to see you back here someday."

"It's a pity you got 17," Snedden added. "But keep working hard. Do what you did here, and you never know, someone might see it and say something to the chief."

"Thank you," I said. They stood up.

"Let's see how your family's doing," said Snedden.

On the way home, I asked my wife what the guys had talked to her about in the kitchen.

"Oh, they asked me about you," she replied.

"Like what?" I asked.

"They said, 'Is he like this at home?' "

"Like what?" I said again.

"That's what I asked," she said.

"And what did they say?"

"They said you never stop."

"And what did you say?"

"I told them you were pretty tired when you got home. They said, 'We never see it.' And I told them you are pretty amazing," she added.

"I'm not sure what kind of impression I made," I said. She looked over at me affectionately.

"I can tell you earned their respect," she said.

6

JIHAD

First impressions can be deceptive. When I first met Kyle Watson, I remember thinking, "what a genuinely nice person." Affable, well-spoken, and humorous, he made me feel at home in my new station. I had felt some trepidation upon being assigned to Station 17. Who would the seven people be with whom I would rub shoulders for the next several years? When Kyle met me on the bay floor that first morning, I had an impression of a lanky, laid-back individual with large ears. He helped me set up my gear on Ladder 17, explained the daily routines, and showed me to my locker. Throughout the day, we exchanged small talk and felt out what we had in common. It turned out we were both married with two daughters, had grown up working on farms, and of course both shared a passion for fighting fires.

17 was a newly built station in the South End Suburbs. It housed an engine, a ladder, and a water-rescue pickup with a motorboat in tow. The district was part urban, part rural, and new subdivisions were springing up every month in what recently had been farmers' fields. My new crew seemed agreeable upon first acquaintance. Captain Peters was a slight, soft-spoken man, the lieutenant seemed friendly enough, and the remaining five firefighters greeted me with warm handshakes all around. I believed that I would settle in well and find my place.

My first friction with Kyle was when I mentioned to him

that I was taking my family to church the next day. Being Irish Catholic was fairly conventional, I thought, and so I was surprised when this information brought out another side of his character.

"How can you believe all that shit?" he asked, his eyes hard. "I mean, it's okay if you want to believe it, but I got a big problem with people teaching it to their kids. You're brainwashing them."

I swallowed, not sure where to begin, but he wasn't finished.

"You gotta let your kids decide for themselves. I wouldn't make my kids believe religion. I'd wait till they're old enough, and let them decide for themselves."

Here was a problem. How was I to explain about the duties of parents and the rights of children to a religious education? I tried an angle.

"You wouldn't deprive your kids of food, or air, would you?" I asked. "If I don't give my children what is essential for their growth, it would be bad for them."

His response was loud and scornful.

"Yeah, but no one needs religion! People need food and air, but religion is just extra."

I sighed. "That's where we differ, Kyle. I think belief in God is essential. You don't. I guess we'll have to agree to disagree."

"Yeah, well, go home and think about what I've said," he sneered. "Cause I won't lose any sleep over it."

So began a relationship that was to make my life very painful for the next three and a half years. Despite his lovable characteristics, Kyle's blind spot was a deep antipathy to anything religious. He would often bring up some item in the news that cast people of faith in a bad light and angrily demand an explanation. Regardless of what denomination was involved, he seemed to hold me personally responsible. He didn't know the differences between the main branches of Christianity, or even, I suspected, any of the major world religions. In any case, I had somehow become implicated in the flaws of all of them.

"I really don't know why Scientologists do that," I would patiently explain. "All I know is that the Catholic Church handles her finances quite differently." Or,

"No, I don't agree that people should set off bombs in the name of God. It's actually against my church's teachings to do that." Or,

"Yes, I agree that Flat-Earthers are crazy. Please don't lump me together with them."

"Well it's all the same!" he would inevitably finish with. "Religion! It ought to be abolished!"

At that time, the sex-abuse scandal was rocking the Catholic Church, and in the minds of many of my colleagues, the word "priest" had become synonymous with corruption. Kyle, of course, could barely contain his scorn.

"You people!" he bellowed at me one day. "You harbor criminals! How can you sit there and believe all that shit when your leaders are child-molesters?"

"You can't paint all priests with the same brush," I replied, fighting back my anger at his unwarranted aggression.

"Yeah, you can!" was his retort.

"Look," I said, "for every bad priest, there are a hundred more doing their best to live good lives. I agree that what these men did was terrible. And believe me, they're being weeded out and put in jail, where they belong. But you have to remember that they are living directly contrary to the principles of the Church. So to blame the Church for what they did is wrong."

None of my arguments made any impression, and he continued to plague me with disparaging remarks.

He was not all hostility, however. Sometimes a month would go by without anything being said on the subject, and he would act companionable enough. One day, I happened to mention that I was looking for a snow blower. My driveway was 900 feet long, and as our first winter on the new farm drew near, my wife and I were scrambling for a snow-clearing solution. The next shift, Kyle called me over and led me out into the parking lot. He opened the trunk of his van, and there inside was an old snow blower.

"I don't use it any more," he said. "Runs great. If you want it, it's yours."

Another time it was a miniature piano.

"My girls have outgrown it," he told me. "I thought yours might like it."

Despite these sporadic gestures of friendship, the attacks continued to surface at intervals. One of Kyle's favorite games was to pretend to be interested in hearing about my beliefs. He would wait until the crew was assembled, then ask me a seemingly innocent question. Sometimes it was to do with a controversial teaching of the Catholic Church, or a Bible reference needing clarification. I would naively launch into an explanation, only to have him cut me off in mid-sentence with a derisive remark. This never failed to get a laugh from the others. My officers seemed disinclined to put a stop to the nonsense, so I resigned myself to the belief that this was simply life in a fire station. In hindsight, I would have ended the trouble much sooner. In those days, rookies were told to keep their heads down and their mouths shut. "A thick skin and a sense of humor are essential for survival in the fire hall," our training officers told us. And so, I blindly obeyed this advice, feeling a rising tide of resentment and a growing conviction that no human skin, however thick, can shield one from the pain inflicted by prejudice.

Things came to a head one Sunday in the kitchen. He was going off on one of his tirades, this time about the Sacrament of Confession.

"You mean to tell me," he roared, "that if a murderer comes to a priest and confesses, the priest won't call the police?"

We had been engaged in debate for almost an hour. He was aggressive and intransigent as always; I was trying with my usual misplaced patience to give some perspective. The altercation was fortunately cut short by the tones, and we ran for the truck.

We could see the smoke as soon as we pulled out of the station. Beyond the rooftops to the northeast, a heavy black pall was drifting slowly skyward. Kyle was at the wheel, weaving the heavy ladder truck expertly through the streets. With every turn, the sheer size of the fire was becoming apparent. We sped out of the suburbs and down a country road. Kyle slammed on the brakes as another fire truck pulled out of a side road in front of us. With intersections and traffic now left behind, the two trucks raced forward at full speed. We rounded a bend. A factory farm lay in full view, gigantic balls of flame rolling upward from what had been a barn. The smoke plume was simply colossal, towering hundreds of feet above us and filling the barnyard, the roadway, and the surrounding fields with choking vapor. Kyle parked on the side of the road, and, gathering our tools in haste, we ran across the pavement just as the crew from the first arriving engine was pulling off their line.

"We need big water!" their captain shouted. His eyes were wide.

"We'll get another line," our lieutenant shouted back. "Grab the long lay!" he ordered me. I raced to the engine and began pulling off the 400-foot length that was packed tightly at the rear of the truck. Kyle hurried to help me, and together we muscled the heavy load onto our shoulders, dividing the weight between us. There was the buzz of a small engine. The lieutenant had got the rotary saw out, cutting away the page-wire fence that lay between us and the fire.

"Go!" he shouted, as the last strands fell away. We advanced into the field. A figure loomed out of the smoke, jacket open, white helmet askew. It was the District Chief.

"Save the farmhouse!" he yelled at us. "Get between it and the fire!" I lunged forward, Kyle following.

~

It was some hours later. Grimy, exhausted, and soaked with sweat, we stood back and took in the scene. The farmhouse had been saved, but the barn lay in twisted ruins. Beneath the rubble, still smoking, lay the remains of eighty milk cows. A heavy stench hung in the air.

"How are you doing?" Kyle asked.

"Good," I said with a weary grin. "This is what I signed up for."

It was another two hours until we got back to the station. Kyle seemed subdued, and we went about our cleanup duties without saying much to each other. I showered and changed into a fresh uniform. We met up again in the kitchen. He was cutting onions.

"Sorry about earlier," he said, glancing at me briefly. It was an opening.

"We've talked religion . . . a . . . a few times," I began, haltingly. "And you always seem so angry."

He looked at me with a puzzled expression. I paused, too, uncertain of how to proceed. It was the first time I had tried telling him about how his behavior was really affecting me.

"There's nothing I can say," I ventured at last, "that seems to be able to break through that wall."

He looked at me with his head on one side. I could see the gears turning slowly.

"Huh, I never really thought of it that way before," he said finally. "I guess, I just find the way you think really scary. I mean, you seem so brainwashed. Always tough to see the other side, I guess."

It was to be the last time we ever discussed religion. He seemed to steer away from the topic after that, and I could only guess whether his contempt for my values was still alive or not.

Looking back, what I experienced was a classic case of workplace harassment. I often struggled during those years to define what was happening to me. My instinct was to engage in the debate in the hopes of planting a seed, hoping that someday it might lead to his conversion. However, the basic respect necessary for such conversations to be fruitful was sorely lacking. Instead, I put up with a lot of abuse in what I thought was a bid to win his soul. Only in hindsight did it come clear that I should have put a stop to the harassment first. If I had showed some spine, perhaps he would have respected me enough to listen to my side of things. Unfortunately, we never got to that point.

It wasn't until some years later, when I was reading through an officer development manual, that I gained a more mature understanding of conflict resolution. According to my training, human beings tend to follow three predictable patterns when subjected to bullying. The first is the Passive Approach. People who are naturally shy, or who lack basic self-confidence, often default to this approach. Passive people, such as I was throughout most of my time at 17, try to shrug off the behavior and be friendly to the aggressor in hopes that it will go away. They usually experience a buildup of unresolved frustration that can suddenly vent itself in inappropriate and damaging ways.

The opposite solution is the Aggressive Approach. This is when a person who has been wronged reacts with extreme anger. I had often heard stories of this happening in fire stations, and even witnessed a few incidents myself. Firefighters tend to be very passionate people, the classic "Type A" personality, and quarrels of this sort can be fairly common. The exaggerated response quells the bully but leaves a damaged relationship in its wake. The two are seldom on speaking terms afterward.

The third way is the most difficult. It is called the Assertive Approach. It takes the most courage because it involves confronting the perpetrator in cold blood. The victim pulls the bully aside and calmly but firmly states his case, demanding that the harassment cease. If the bully reacts unfavorably, the victim simply asserts that he will be taking the matter up the chain of command. It is as simple as that. Either you treat me properly, or the game is up. Working oneself up to that moment can be profoundly difficult, as one may have to overcome months if not years of conditioning to being dominated by that person. Much of what I endured could have been stopped, or at least reduced, had I known this simple but powerful formula.

Even without that knowledge, I still had a peace that came from my own convictions. I had grown up absorbing lessons about the importance of forgiveness, and in my difficult years at 17, I was given an opportunity to put those lessons into practice. Naturally, my spirit revolted a little at the idea of forgiving Kyle, but after the first effort came the freedom that accompanies mercy. It was a real-life lesson in the power of forgiving one's enemies. Confronting the aggressor is important, I realized, but forgiving them is even more so. It is the only way to find true peace. I often prayed for him in later years, hoping that something I had said would eventually take root. I never did learn whether his attitude changed, and after being transferred to a different station I never saw him again. I could only hold him up to God's mercy and surrender the results of my halting efforts to His all-seeing providence.

7

WAR WOUNDS

"There's no sense getting your nose out of joint over this one," I said. The elderly nurse paused her assessment of my bloody face and laughed.

"Well you must be doing all right if you're still cracking jokes," she said.

I was sitting in an examination room at the Central Hospital. A chunk of a car door hinge was lodged firmly into the bridge of my nose, and the nurse, my lieutenant, and a safety officer were gathered around forming a concerned audience.

"Tell me how this happened," she said in a reproving but motherly way.

The day had started like any other. We had finished our morning checks and were gathered around the long kitchen table to go over the day's routines.

"We need to order station supplies today," Captain Peters read off the list in his quiet voice. "Ben, can you take care of that?"

"Sure," I replied. As the junior man, this task often fell to me.

"Kyle, you can show him how to print off the list," he went on. "We'll get the station cleaning done early, then we're on for extrication training at 9 o'clock."

Extrication day was an important event. Engine 17 was a rescue pump, which meant that in addition to carrying standard medical and firefighting gear, it was equipped to respond to serious car accidents. We carried powerful hydraulic tools on board, which could shear through metal, pop doors off their hinges, and lift dashboards. The spreaders, sometimes known as the "jaws of life", were my favorite tool. Built like an enormous pair of blunt scissors, they opened outward to force various components of a vehicle apart. Once a month, we would drive over to the local junkyard and spend the whole morning cutting up cars. This particular shift I was assigned to the back of Ladder 17. Both trucks were to participate in the training, and as we roared noisily down the road, I looked forward to sharpening my skills.

We pulled into Bennett's Auto Recycling and came to a halt beside a massive pile of scrap steel. Firefighters climbed leisurely out of the cabs, pulling on gloves and doing up helmet straps. A large forklift drove up beside us and the driver rolled down his window.

"Can you bring us something with four doors, please?" Peters called up to him. The driver grinned, waved briefly and drove off toward the row of wrecked vehicles on the other side of the yard. In a minute he was back, with the crumpled remains of a Pontiac Grand Am on his forks. There was a crash, as the car slid to the ground and sat heavily on bare rims. We gathered around for our briefing, and the captain laid out the training scenario.

"We're pulling up on the scene of a single-vehicle MVC," he explained. "We have one patient trapped in the driver's seat. Ben, what's the first thing we do?"

As the newest firefighter, questions testing my knowledge were a daily rite of passage.

"Assess the scene," I replied quickly.

"Good. What are you looking for?"

"Well," I ventured, "there could be fuel spills, downed wires, smoke coming from the car . . . things like that."

"Right," he said, "and you might want to think about people being ejected. I always take a look in the ditch in case anyone went through the windshield. As the officer, I'll be walking around the accident scene coming up with a plan. You've assessed the scene as well. What's your next job?"

"Making contact with the patient."

"Before that. You can't get into the car until you've done what?"

"Stabilized it."

"That's right. So go ahead and stabilize the vehicle. Show me what you would do. The rest of you guys can lend a hand."

I hurried over to the truck, opened the extrication compartment, and began pulling out the large chunks of hardwood we used to wedge under vehicles. The others crowded around to join me, and together we worked to prep the old Pontiac. The wood was placed at key points under the axles, while Engine 17's driver ferried up the portable motor to power the tools. More hands hooked up the hydraulic hoses and laid the massive spreaders and shears in readiness next to the car. I felt the driver's hand on my shoulder.

"An extrication is like an operation," he said sternly. Graham Dykstra was the senior man. He was burly, prone to anger, and wore a grim expression. "When the crew knows their job," he continued, "it's like a doctor and his team working on a patient. The doctor should be able to hold out his hand, and the nurse should know exactly what tool to put into it. You need to understand the flow and be able to see what has to happen next. There's something comforting in the knowledge that the guy eating spaghetti next to you at the fire station table knows his part. You want to be that guy."

We watched as the engine crew clustered around the car, breaking glass and ripping away the interior trim to expose any wires or airbags.

"Ready?" asked Graham. "You're on the spreaders. Take that door."

I picked up the heavy tool and tried to insert the tips into the tiny crack between the door and the frame. I gave a bit of throttle to spread the tips, and instantly the tool popped back out. Graham was looking over my shoulder.

"Not like that, you'll never get it to spread that way. Crush it first and make a purchase point. That's it, now you have a gap. Now get your tips in. There, now spread. Stop! Close the jaws and drop them lower into the gap you just made. Now spread! Keep going . . . keep going . . . Stop! It's tearing. Now, see the bolt and latch? Get your tips right above it. No, not like that! Angle it up so you can get them right in there! Yeah, now spread."

With a satisfying pop, the bolt broke and the door flew open.

"Okay," said Graham. "Now take it off its hinges." This will be easy, I thought. A nice, right-angled spread. I placed the tool into the space between the hinge and the door. I heard the groan of metal under stress. Suddenly, there was a noise like a gunshot, and something hit my face with tremendous force. I froze, stunned. Then I looked down and saw a steady trickle of blood running down my nose and spattering the ground.

"Everybody freeze!" Graham bellowed. "Ben, put the tool down! Step over here!"

I obeyed mechanically and tried to grin as the crew gathered around curiously.

"Here," said Kyle, "Let me see if I can pull it out."

At his words, I noticed for the first time that whatever had hit me had not bounced off but was actually lodged in my nose. He gave a gentle tug, there was a sharp twinge, and my hand flew up instinctively.

"Nope!" I said quickly.

The lieutenant sighed. "All right," he said. "Let's take you to the hospital."

It was a twenty-minute drive downtown, and as the truck clunked ponderously over each pothole, I kept my hand cupped over my nose to reduce the impacts. We parked in front of the emergency ward and, leaving the driver resigned to hours of waiting, walked through the double doors. The triage nurse didn't bother to look up as I approached the desk.

"Hi, how can I help you?" she droned, her fingers ticking away on a computer keyboard.

"I injured my nose," I said. She stared at her screen, unmoved.

"Do you have your card with you?" she asked, still not looking up.

"Yes," I said, pulling out my wallet. I placed the card on the desk, and she glanced at it, typing.

"Take a seat, please," she said.

"Um," I ventured, "I think I need to see someone right away."

Finally she looked up, and I saw her eyes widen.

"Oh!" she exclaimed.

Moments later, I was sitting on the edge of a hospital bed, the kindly old nurse examining the damage.

"Well, it looks nasty," she said when I had finished my story, "but the good news is it's only being held on by a very small barb. The doctor should be able to get it out quite easily."

The doctor turned out to be a trim young man in green

scrubs. He peered at the protruding chunk of metal for a moment, then reached out and gripped it firmly between his thumb and forefinger.

"Hold still," he said. Before I could say anything, there was a sharp tug, and the piece came off in his hand.

"There," he said soothingly, "all better."

It was three hours later, and I was almost home. I let my pickup roll slowly down the long driveway, savoring the view of the hay fields, the old gray barn, and the little farmhouse where my family was waiting. What a surprise for them, I thought, to have me home early, with a week off thrown into the bargain. The safety officer had been adamant that I be released on sick leave for at least that long, and after drowning me in paperwork, he had sent me packing. Kate greeted me at the door, holding the baby in one arm. I kissed her and opened my arms wide to greet the two little girls who ran up with joyful cries of "Daddy's home!" Suddenly my wife stepped back.

"What's that on your nose?" she asked.

"A band-aid," I replied, feeling the 1-inch strip which was all I had to show for my mishap. "I had a little accident at work," I explained. "Nothing serious, and the good news is, I'm home for a whole week."

The little ones clapped their hands in delight, and my wife's blue eyes sparkled with humor.

"A whole week?" she said. "For *that*?"

8

SPARK PLUG

Marcus Lambert's look didn't suit him at all. Long hair on a fireman was unusual enough, but when I first caught sight of him rounding the corner on his bicycle, I thought it was a homeless person stopping by the station. His silver pony tail was tied up in a yellow bandanna, a scraggly moustache adorned his upper lip, and he was dressed in faded sweat clothes. It wasn't until I saw the standard issue gear bag on his back that I realized it was Marcus, arriving for his first day as acting Lieutenant.

We ran a few inconsequential calls that morning, and as I drove he kept up a series of dry and sometimes cutting remarks.

"What, you didn't see that pothole? You only had five hundred feet to slow down for it."

Or, "Why didn't you turn around back there? Now we gotta cost the city another fifty bucks in fuel to go around the block!"

He was not all criticism, though, for on the next alarm call I pushed the 43-foot, 65-ton ladder truck to its limit, and we beat Engine 16 in by a good two minutes. Upon arrival, Marcus gave a rather unorthodox radio report.

"Dispatch, Ladder 17 on scene, already . . . because Ladder 17 has a *superior* driver."

He threw down the microphone, laughing.

"That comment will earn me a phone call from the chief!"

Later that day, we were chatting in the station lounge, and it came up that my wife and I were putting in a garden.

"How big of a garden?" he asked.

"I'm not sure yet," I replied. "It'll depend if I can get a rototiller or not."

"I got one," he said. "Actually, I was planning to sell it."

"Oh, yeah?" I said, interested. "Would you mind if I come by and take a look?"

"You give me a ride home after work, and I'll show it to you," was the reply. "It's a solid little machine. Strong, too. It could bust up concrete."

After shift change, I met him in the parking lot and loaded his bike into the back of my pickup. Marcus lived only a few blocks away, and we pulled up in front of a stately town house with a well-kept lawn.

"Come on through," he said. "It's in the back."

He led the way into the yard, past a small vegetable plot, and opened the door to an old tool shed.

"There she is," he said proudly. I looked in. An ancient machine lay in the shadows, leaning drunkenly on a flat tire. At some point in its life, it must have been red, but the paint had faded to a sickly orange-brown, mingled with rust spots.

"I see," I said. "Does it run?"

"Of course it runs!" he exclaimed. "It's always been reliable. I used it for years."

"How come you're getting rid of it?"

"Just picked up a new one. Not that there's anything wrong with this one," he added.

"Can we start it up?"

"Not here. I don't want to bother the neighbors. If you can put it in your truck, we'll go over to the strip mall and fire it up in the parking lot."

We pulled the machine out into the sunlight and began wheeling it across the lawn. It was surprisingly heavy, and the damaged wheel dug into the grass, leaving a streak of ruin behind it. We paused for breath, and something seemed to occur to Marcus. He disappeared into the garage and came back carrying a small air compressor.

"Let's pump that up," he said. He attached the nozzle to the valve stem, and I watched as the air hissed into the tire. Several minutes passed, and it showed no signs of inflating. Finally he yanked it out.

"Didn't think it would work," he mumbled. "You'll probably have to replace the whole wheel."

We left a furrow as we hauled it slowly around to the front of the house. It took all our lifting power to get it into the back of the truck. I drove around the block and parked in an empty lot behind the liquor store. We got out of the cab and climbed into the back.

"Good thing I remembered this," said Marcus, pulling a small object from his pocket. Looking closely, I saw that it was a spark plug.

"I use this on all my machines," he said. "Came off the tiller originally, but it fits my lawn mower. Why buy two spark plugs when one will do? I don't cut the grass and till the garden at the same time!"

He screwed it in, and producing a small wrench from his pocket, he tightened it snug.

"Now," he said, "let's fire her up."

He grasped the handle of the pull-cord and gave a yank. There was a whirring sound, then silence. He pulled again. Nothing.

"Come on," he growled and began hauling repeatedly with all his might. The whirring sound continued, but the motor refused to start.

"You try!" he ordered. I gave a few tugs, but it was clear that we were not going to get anywhere. I straightened my back.

"Is there gas in it?" I asked. He looked at me for a moment, then bent and unscrewed the cap. The tank was bone dry.

"I don't know if I really . . . " I began, but he cut me off hastily.

"Don't worry," he said, "we'll get it going." He jumped out, hurried to the cab, and produced a small jerry can.

"I think of everything," he said triumphantly. He inserted the spout and poured in about a tablespoon.

"There," he said, "that's all you get. Gas is expensive these days."

He unscrewed the spark plug, poured a drop into the socket, then replaced it.

"That's a good trick to know," he said. "If anyone asks, I taught you that. And by the way, never touch the top of that plug while the motor's running. You'll get a shock."

"If we ever get it running," I muttered.

He pulled the cord again, and this time to my surprise the machine sputtered into life.

"There!" he cried. "Didn't I tell you!"

We let it run for a few minutes, then shut it off. I was beginning to realize that I would be inheriting a project.

"Okay," I said slowly. "How much were you thinking?"

He inhaled through his teeth and blew out thoughtfully through his moustache.

"I paid the guy seventy-five for it," he said. "Mind you that was fifteen years ago. Then I put a new belt in. I've probably sunk a hundred bucks into it all told. But, I'd settle to get my seventy-five back."

I pulled out my wallet and looked inside.

"I've got forty dollars," I offered. "Would you take that?"

"The old that's-all-I've-got-in-my-wallet trick, eh?" he snorted. "Tell you what. I'll take the forty, and you buy me lunch some time, how's that?"

"Agreed," I said. We shook hands.

"You can take me home now," he said. Then he pulled out the little wrench again. "I need my spark plug back though."

As we drove, he revealed a little more about my purchase.

"You'll have to figure some way to get that wheel off," he said. "Maybe a torch. I've tried, but I could never get it off. I think it's welded to the axle. That's your problem now."

I dropped him off and headed home, not sure how I would explain it all to my wife. I had told her over the phone that we were getting a good deal. It now seemed that the "deal" was more of a headache. She accepted my blunder with good grace, however.

"At least," she said, "it only cost us forty dollars."

Marcus was off the next shift, and I told the story at the table to the amusement of my colleagues.

"Typical Marcus," the captain shook his head. "You're lucky you fared out as well as you did."

It was to be a lesson in the importance of refraining from gossip, for the next time I saw him, Marcus came up to me with a frown of irritation.

"I got a call from Jeff Fraser over at Station 16," he said. "He told me he heard that I'd screwed you over on a deal."

"I never said that," I said quickly. "I was just telling the guys that I'd bought a tiller off you."

"Yeah, well, you gotta be careful what you say around here," he lectured. "Word gets around."

"I'm sorry," I said innocently. "I meant no harm. I just happened to mention that you needed your spark plug back. For some reason they thought that was really funny."

Today, the rototiller is still sitting in my garage, leaning drunkenly on its side. I never have succeeded in getting that wheel off.

9

ROUGH WATERS

Graham Dykstra had no patience for rookies. Fresh from Station 3, I was all determined to make a good showing of my work ethic. One of my first projects was to wash the station windows. I set to my task whistling, conscious that this had been a proven method of earning the downtown crew's respect. It caught me by surprise, therefore, when a door banged open and an angry face appeared.

"What are you doing?" Graham demanded loudly.

"Washing windows," I replied. "I had a little spare time, thought I'd make myself useful."

The frown deepened. "Well you don't need to be doing that stuff all the time. If the windows need cleaning, we'll do it as a team."

The door slammed again, and he was gone. It was the first of many such interactions. Graham seemed to believe from the outset that I needed putting in my place, and he didn't bother to use much common courtesy in his speech.

One task I had been used to doing on probation was taking care of the grounds. At the Big House, I had been ordered to pull the weeds out of the cracks between the paving stones. 17's walkway was in a sorry state, so I dutifully began what I thought was expected of me, yanking up the crab grass and chick weed. Again the door banged, and again the angry head appeared.

"Get out of there! You should be learning firefighting, not gardening!"

This time, though, I noted a glint of humor in his eye and perhaps even the hint of a smile at the corner of his mouth.

He's just yanking my chains, I thought. Still, it wasn't the nicest style of joke. Later, we were practicing hose advancements on the bay floor. The captain wanted each firefighter to pull off the 200-foot attack line and drag it through the double doors that led into the living area, stretching it down the hallway to a predetermined target. The point of the drill was to develop a sense of how far one could go with that much hose as well as to get comfortable with the mechanics of deployment. When my turn came, I shouldered the heavy load, jogged away from the truck, and passed through the doors and into the hallway. I felt a tug as I came to the end of the 200 feet. The captain nodded in approval and ordered me to reload the hose.

"I'll get you to do it one more time," he said, "just for practice."

Graham had not overheard this comment, and he helped me replace the attack line without saying much. He seemed to think that the training session was over, for as soon as we were finished, he began removing his helmet and coat. Remembering my officer's orders, however, I pulled off the line again, came around the back of the truck and headed for the doors. I had just passed through them when there was an explosion of fury behind me.

"Stop!" It was Graham's voice. "What are you doing? Get back in here, you idiot!"

I stopped and turned to face him, feeling the color rise in my face. There was no trace of humor on his. The captain was out of sight on the other side of the truck and either hadn't noticed or was unconcerned about Graham's aggression.

"You don't need to do it again," Dykstra said, lowering

his voice to a more conversational tone. "That's our living quarters. We shouldn't be bringing a hose in there in the first place. Contaminants and all that. You can go ahead and repack the hose."

I swallowed my pride and began cleaning up in silence. It had only been a week, and so far things didn't seem to be going well with my co-worker.

Despite these outbursts of temper, there was one side of Graham that I was grateful for. He had been raised in a mainline Protestant denomination, and though he was not a regular churchgoer, he seemed to understand where I was coming from on religious issues. Whenever Kyle was giving me a hard time, he would back me up, often articulating things better than I could. Where Kyle was disdainful, Graham was generally respectful, even though he didn't necessarily agree with everything I believed. When it came to firefighting, however, he remained my toughest critic. Kyle, on the other hand, took my rookie mistakes in stride, and if Graham was overly harsh he would say:

"Don't worry about him. He likes you, he's just very particular, that's all."

Number 17 had some differences in culture from my previous stations. In the months that followed, I was forced to unlearn a lot of the expectations that had become ingrained on probation. One of these was fighting for dishes. One morning, something possessed Graham to beat me to the sink. I maneuvered into position beside him and applied my standard tactics.

"Let me in there!" I said cheerfully. He turned and glowered at me.

"Don't. I'll be seriously annoyed," he growled. Not to be blustered, I dug an elbow into his ribs.

"Get outta here!" I ordered. He threw down the dishcloth

and stalked out of the kitchen. Taken aback by the seriousness of this reaction, I watched him go with a tightening in my throat. It seemed the unwritten rule of cordiality that governed battles over dishes was lost on him.

These were minor skirmishes compared with what was to come. The following summer, things came to a head in an incident that would forever live in my memory. 17, being a water rescue station, had a 16-foot, rubber-hulled boat with an outboard motor. As soon as the ice came off the river, we conducted regular training exercises. The stretch of water in our district was relatively calm, but the captain liked us to practice white-water tactics in the event that any of us were sent to a different station. I had never had much experience with power boats, and this particular day I was to be shown the proper technique for picking a victim out of rapids. One firefighter would stay in the water, dressed in a wet suit, life jacket, and rafting helmet, while the boat approached from downstream. At just the right moment, the operator had to kill the motor and spin the wheel so that the stern of the boat, with its deadly prop still spinning under the force of the current, moved away from the victim. If done right, this would cause the boat's nose to swing around between the victim and the current, placing them within arm's reach of the rescuer crouching in the bows. Timing and precision were everything.

When my turn came, the victim was none other than Graham. Easing the throttle forward, I pointed the boat toward the bobbing head in the water. I was unfamiliar with that stretch of river, and without realizing it, I was approaching him, not from below, as was expected, but from upstream. I slowed, but not soon enough. I realized with a spurt of panic that I was coming in too fast. I had a last sight of

Graham flailing to get out of the way before the boat swept by him with inches to spare. I remembered the prop and turned the wheel to move the stern away from him, but he still got a mouthful of wash. He sputtered, his face purple with rage. Kyle, beside me, shook his head.

"Way too fast, man. Try it again. This time come from downstream."

We circled around, and I aimed for the figure in the water, this time at a snail's pace. Creeping up, I swung around Graham, but now I was too cautious, and the boat stayed a good two feet out of reach. I felt a tap on the shoulder.

"Here, let me demonstrate," said Kyle. I surrendered the driver's seat with some embarrassment and watched as Kyle and his lookout man in the bows pulled off a textbook rescue. Graham was hauled into the boat and sat puffing and red-faced as the water dripped off him. When he was ready to speak he turned on me.

"You could've killed me!" he snapped. "You gotta pay attention! I don't know what's going on in your head, but you clearly don't have a clue what you're doing! If you wanna drive this thing, you better get your head in the game!"

He retreated into sullen silence and sat huddled in the prow during our journey back to the dock, while Kyle cast a sympathetic glance my way, and the captain admired the passing scenery with a complacent expression. Little was said as we loaded the boat onto the trailer and drove back to the station. We cleaned up, and the crew seemed to scatter. I went to my locker and changed into my gym clothes. As I was starting my afternoon workout, I heard my name called over the station speaker. I walked into the kitchen and saw the whole crew assembled at the table, regarding me with serious expressions.

"Take a seat," said the captain quietly. With a tightening of my stomach, I sat down to face them.

"First of all," said Peters, "how did that go today?"

"Not good," I replied. It was never a good thing when someone asked "how did that go?" It was always a lead-up to a reprimand. The lieutenant was the next to speak. I knew he was tight with Graham, and I could tell from his face that I had earned his displeasure.

"When your senior man tells you to do something," he said sternly, "do it the way he says. Don't keep doing the wrong thing."

"Okay," I faltered, not sure what to say. I knew I had messed up, but it had not been deliberate. Graham cut in.

"We told you THREE . . . TIMES to come from downstream," he said harshly. I only remembered one time, but I didn't get a chance to say so, for he charged on. "If that's the way you think it should be done, then you and I have some serious issues."

"I'm sorry," I said, "I wasn't trying to run you over."

Kyle forestalled what was shaping up to be an explosive reply.

"We just want to make sure you're on the same page as everyone else," he said, not unkindly. He proceeded to explain the proper technique in more detail, while I strained to listen, conscious of my hostile audience.

"I get it now," I said, when he had finished. "And I wasn't ignoring anyone on purpose. I guess I just got a little turned around. It was flat water, and I got confused which way was upstream."

Graham looked thoughtful at these words, and said in a less hostile tone: "I know it wasn't on purpose. Not everyone is at home on the water. It's not our natural environment.

Maybe we need to take you out a few times and practice recognizing current. If you're not sure about something, just speak up next time."

"Yeah, and we're not trying to gang up on you," the lieutenant put in. *Oh, no?* I thought.

"No," said Graham. "But there are some issues that need to be addressed. First of all, do you not like us or something? You never talk. Everyone else talks a lot here except you."

It dawned on me. This was about more than just the boat. I realized with a sickening feeling that he had succeeded in turning the whole station against me.

"Look around you," he continued. "Firefighters are a loud bunch. We like to tell stories and make fun of each other. You don't have to say anything that goes against your deeply held beliefs, but we want to see you at least try to be part of the team."

I felt it was time to say something in my own defense.

"It has nothing to do with not liking you guys," I said. "I was told in drill school that as a rookie I should keep my mouth shut. That's all I'm doing."

They digested this momentarily.

"Yeah, but you're past that now," the lieutenant said finally. "You're what, two years on the job?"

"Three," I answered.

"All right, so you're definitely at the point where you can speak up. Sure, keep your mouth shut during ladder drill, but the rest of the time you can be part of the gang. Tell us about your life, what you did on the weekend."

"I can do that," I said humbly.

"All right, enough of this," he said, slapping his hand on the table. "What's for dinner?"

At last the sun went down on what had been a day of defeats. Glumly, I viewed my options. How could I have gone so far wrong, I wondered? In reviewing the past year, I had to admit that my behavior was not calculated to win friends. Between Kyle's anti-religious onslaughts and Graham's temper, I had kept to myself in my spare time. Added to this, I had often felt compelled to remove myself from conversations that were inappropriate for the workplace. Kyle, in particular, delighted in bragging about his sex life in lurid detail, at which point it was my habit to leave the room quietly. All of this had added up without my knowing to a general bad feeling where I was concerned. I realized that I hadn't exactly put a big effort into being a good brother to them. I toyed with the idea of asking for a transfer, but I knew that if I left now, I would always be remembered as the sad dunce who couldn't drive the boat or fit in. There was nothing for it, then, but to try to be more friendly and to master river rescues.

I set to my new task the very next shift, forcing myself into the role of the extrovert. It was not always easy. My co-workers' interests chiefly revolved around pop culture, TV shows, celebrity gossip, and electronic gadgets, topics of which I had little knowledge or interest. I would often find myself jumping into conversations where I was out of my depth, which gave them fresh ammunition for ridicule. All the while Graham remained aloof. Sometimes he would deign to engage me in brief conversation, but he was never as open with me as with anyone else, and there was always a hint of condescension in everything he said. The boat incident was the last major explosion I ever experienced from him, but a cold wall remained in its wake. As the months wore on, it became more and more difficult to go to work in the mornings.

Despite these challenges, I had some factors in my favor.

The first was Russ Hennessy. Russ was a decorated military veteran who had decided to become a fireman. He had served two tours in Afghanistan and one in Iraq. He swaggered, swore, and had seen things I couldn't even imagine, yet despite his rough exterior, he never had an unkind word for me. We would chat about hunting, sports, and the challenges of raising children. I was on a difficult road, but in him I felt I had at least one ally. A change of officers also helped my situation. Graham's friend the lieutenant was transferred, and Chuck Brown came in to take his place. Chuck was a round, bald-headed sanguine who took an evident liking to me. With his warmth and outspoken confidence in me, I saw light at the end of the tunnel. The third factor involved a change in my own abilities. I was sent on an official course to become a boat rescue technician. After three days of intensive training, I felt supremely confident behind the wheel. Upon return to the station, the captain observed my newfound skill set with approval, but Graham, as was to be expected, said nothing.

I sometimes confided a few details of my trouble to my Dad, asking him for prayers. One day, I got a note from him in the mail. There was no introduction, just these few words:

Hold Your Line

> Satan taunts you, tempting you to step out into the zone between the two battle lines. He makes you think you can defeat him there, by engaging him on several levels of provocation. Remember at all times that you must not do this. Do not let him provoke you to rage (realize that at the root of your rage is your fear). If you leave your station in the battle line, you break your own line and weaken the lines behind you. Stand firm. Hold your position, even if you do not understand its purpose, or its usefulness, so that when

the King tells you what to do, you will be ready to do it and you will be effective. Hold your line! Stand firm! The battle belongs to God.

~

I pulled hard on the cord above my head, and the air horn blared. Cars scuttled out of the way, and my foot found the gas pedal again. Ladder 17 roared through the intersection and on down the street, siren wailing. The March night had turned cold, and the roadway was shiny.

"Left at the next lights!" Graham called from the back. The traffic signal ahead had turned green, and I could see its squiggly reflection on the icy roadway. My foot went to the brake, but something didn't feel right. We were still moving too fast. Now we were over the stop line; it was time to turn the wheel. I spun it hard to the left, feeling the back end slide sideways at the same time. My foot came off the brake. The truck came out of the skid as if by magic, and we were in the right lane, facing the right way. I breathed a sigh of relief and let the truck slow. I caught my officer's eye. The look spoke volumes.

"Take it easy, buddy!" It was Graham, speaking with his habitual hardness. The radio cut in.

"Ladder 17 from command. This is a confirmed false alarm. You can return."

Back at the station, Graham let me know how he felt in his usual cutting way.

"Slow down when you're coming up to an intersection," he snarled. "I thought we were going to slide off the road."

He was right, but his manner, coming on top of so many months of ill-treatment, was simply too much. Something had to give. I had endured him in silence long enough. I had not yet built up sufficient courage to sit him down for the much-needed one-on-one, but I knew it must be done. I fully expected him to react unfavorably. I had seen him turn too many arguments around on his opponents to anticipate anything less. Still, I hoped, if I could make my case strongly enough, perhaps down the road there would be a lessening of aggression.

My chance came sooner than expected. At the start of the next shift, I arrived at the cab door to find the lieutenant had not yet arrived and that Graham was already on board, checking his breathing apparatus.

"Morning," I said.

"Morning," he returned. "How's it going?"

"Not very good," I said. He stopped what he was doing and looked up.

"Why? What's up?"

"This job is keeping me awake at night," I replied. He turned to face me. I had his full attention now.

"That doesn't sound good. What's causing that?"

"Well the driving for one thing. Like what happened last shift."

"I wasn't mad at you," he said quickly. "Don't take that stuff personally. Just accept where you're at. Eventually it's going to click, and one day you'll say to yourself, 'I'm good at this.'"

"It's more than that, though," I said. "I feel like I've made you mad at me a lot."

His next words took me completely by surprise.

"You're a good fireman, Ben. You're a very good fireman. You know what I see in you? You're dependable. You get the job done. No one's expecting you to be perfect. But I know the captain likes having you around. You take advice. You've gotten a lot better at being part of the conversations."

"Thank you," I said, biting my lower lip to quell the unexpected emotion that was welling up. He went on.

"I know there's a lot of yelling around here, but that's because we have to, because it's dangerous."

I looked down, shaking my head.

"No? You don't agree?"

"I agree that things need to be addressed," I said. "But it's all in how you do it."

He nodded.

"I've had to rethink a lot of things," he said. "Do you remember a while back, we had the boat out, and I was worried you were going to run me over?"

"Eight months ago next Friday," I said dryly, "I remember."

"I was rude to you that day," he said regretfully. "It's kept me awake at night, too."

I looked at him in disbelief.

"It's my first time being a senior man, you know," he explained. "Having to be the one with all the answers. I guess this has been a learning experience for all of us."

"It sure has," I agreed.

It was a reconciliation of sorts. Although we never became friends, there was a lightening of his manner toward me from that day forward. It didn't take me long to realize that I had done all I could at Station 17 and that it was time to move on. I would never fit in here, but at least I had shown that I was able to do the job, and they had seen that I had tried to be a good friend to them. Now it was time to find a station and crew where I could belong. One morning, I walked into the captain's office.

"I'd like a transfer," I said.

"A transfer!" He looked up, surprised. "You don't like water rescue?"

"I've got nothing against water rescue," I replied, "but I'm ready to move on. I think I'd like to try a busier station."

"Well, that's fair. I suppose you're young, and you want to get in on some action. You write down some stations you'd be interested in, and I'll send an email to the chief. Transfers are coming up in a couple of months."

I left his office, feeling like I was walking on air. I had slogged through three and a half years as the black sheep of the station, endured the bullies, and conquered the boat. I had done my time, and soon I would be free.

Some years later, I decided to write to Graham. I was nervous about revisiting something that he would probably consider resolved, yet I felt that I had never fully communicated the impact of his behavior. In a few short lines I told him how hurtful his aggression had been and expressed my desire for a more respectful work environment, should we ever be stationed together again. A week later, the phone rang.

"Ben," it was Graham's voice on the other end.

"Hi, Graham," I said. "You got my letter?"

"Yeah," he said slowly. There was a pause. "I was . . . very surprised," he went on. His voice was more subdued than normal. Another pause. "I don't really know what to say," he finished. I gathered my thoughts. I hadn't exactly expected an apology, but I knew we had to talk this out.

"I needed to get it off my chest," I said simply. "I've had trouble communicating with you in the past, and I felt it was time to deal with it."

There was silence on the other end, then: "What do you want to come from this?"

I could sense the unspoken question behind his words. He was worried about a harassment charge.

"You don't need to do anything, Graham. I just thought you should know, that's all."

"You just want to state your case and walk away from it, type thing?"

"Yeah, I guess so. I was also hoping that if we ever work together again, we could be on better terms."

"All right. Look, I don't want to have an awkward conversation with you or put any more pressure on you. But, just so you know, there really was nothing. I don't *have* anything against you. If you think I'm going around with a chip on my shoulder . . . I'm not."

"Thanks for the reassurance," I said. "I assumed from your manner that you had it out for me."

"No," he said firmly. "I don't mean to minimize what you experienced . . . what it was to you . . . but frankly I was shocked when I read your letter. I had no idea there was any bad blood between us."

"Well," I replied, "maybe it's partly my fault for not communicating in the first place. If I had been like, 'Hey, Graham, it really bothers me when you talk to me like that,' or, 'Hey, let's talk about what happened on that call, you seemed really upset,' it would have been better."

There was another strained silence. I had given him plenty of opportunity to apologize, but he wasn't taking the bait.

"Look," he said finally, "if you're worried that I would have it out for you if we're ever at the same station, I definitely wouldn't. I'm not going around saying bad things about you or looking for an opportunity to cut you down or anything like that."

"All right," I said evenly. "And I want you to know that I respect your experience and would gladly work with you again. I would just hope we could be on better terms next time. If you're happy with that, then I'll leave it at that."

"Okay," he said. "I'm good with that."

"All right," I said.

"All right," he echoed. A final pause.

"Well," he said. "I hope everything works out for you."

"Yep, take care of yourself," I said, and hung up.

I stared into space. Had I misjudged him? I realized for the first time that bitterness and wounded pride had blinded me to the full reality of the situation. Part of me felt annoyed that he hadn't been more apologetic, but the more I turned it over in my mind, the more I realized my own lacking. In nursing my hurt feelings, I had allowed myself to take a one-sided view of him. Certainly he had been rude and unfeeling on occasion, but his was the type of character that charged through life in that way, loud, opinionated, but perhaps more well-meaning than I had been willing to give him credit for. From our conversation, it was clear that he hadn't deliberately played the bully. Yet something about me had decidedly bothered him all those years. My standoffishness, I supposed, and the inevitable mistakes that come with learning the job. I decided it was best to forgive and forget, and resolved to be less thin-skinned next time.

It dawned on me that the greatest trials are not always the dramatic events like the ones that occur at emergency scenes. More often, it is the day-to-day challenges of working with flawed human beings and coming face to face with one's own flaws. Living with the pain of a difficult relationship can be a great burden. Reaching a resolution and obtaining an unexpected insight into another's character and motivation come with a sense of relief. I think both of us learned an important lesson that day: I, not to take myself so seriously, and he, to be aware of how his communication style affected others. Sometimes challenging another makes him realize that his behavior is wrong. It can also reveal his good side and expose one's own faults.

10

SNAKES AND LADDERS

Every so often, I would find myself detailed to another station for the day. This was always a welcome break, and it gave me a chance to meet new crews and see how other stations did things. Any time someone called in sick or went on vacation, the chief would pull an extra from another station to fill his spot. Today it was my turn, and I was headed to Station 16 for the 24-hour shift. This hall boasted a ten-man crew, with an engine, a ladder, a heavy rescue, and a water-rescue unit. I knew there would be plenty to keep me busy.

At twenty after six, I walked in and checked the duty board. I saw my name under "Ladder 16" and headed to the bay floor. The truck was over 40 feet long, with a heavy steel retractable ladder on top and a working platform that rested over the cab. As I was hanging up my coat, I met my officer of the day. He was a friendly, relaxed man who greeted me with a warm handshake.

"James," he introduced himself. "I go by 'Spongy.'"

"Spongy?" I repeated, not sure I'd heard properly.

"That's what they call me," he said, with a smile and a shrug.

We gathered at the long kitchen table, the officers naturally gravitating toward one end, while the "buffaloes" sorted themselves roughly in order of seniority toward the other end. The captain gave a short speech laying out the expectations for the day. He was an animated man with intense eyes.

"We're scheduled to reload hose today," he announced. "We'll pull off the engine's supply hose first. Make sure when you're putting it back that the folds are in different places, and check for rot. When that's done, we'll do some search and rescue drill in the training tower. This morning the heavy rescue is driving down to do a pre-incident plan of the new school. Engine crew can get groceries. Ladder, you guys can stay in station to cover the district. That's all I've got. Thanks."

There was a murmur of conversation and a scraping of chairs. I followed Spongy out to the bay floor to go over the equipment on the ladder truck. There I met my driver, Sam, a solidly built man of few words. I helped him set up the ladder on the tarmac and start the saws. When that was done, he reversed the truck back into the station, and we headed into the living quarters to mop the floors. The morning passed without incident. Station routines were completed, lunch was eaten with lively conversation, free time sped by, and soon we were out on the bay floor again pulling off hose.

Tones. I dropped what I was doing and ran for the truck.

"Engine, Ladder, Rescue. 250 Edwards, between Porter and Sparrow. Smoke visible, inside fire."

Three garage doors opened. Three fire trucks sped out, lights flashing. Then we were off down the street, engine first, ladder following, and the rescue bringing up the rear. Sam was self-possessed in the driver's seat, and Spongy's face looked relaxed but attentive as he listened to the radio reports. I hurried into my coat and breathing apparatus. The call was out of district, and updates from the first-arriving engine were coming over the waves.

"Dispatch, Engine 15 on location. We have a two-story,

detached residential home. Heavy smoke showing from the eaves and around the chimney. Put in a working fire. Engine 15 will be Edwards Command."

My pulse quickened. I was on my way to a fire with an unfamiliar crew in an unfamiliar district. I made a mental checklist: Mask, gloves, helmet, radio, flashlight, axe. We were turning the corner onto the fire street already.

"Set up here," the lieutenant was telling the driver. The truck ground to a halt, and the air brake hissed. My gloved hand found the handle, the cab door flew open, and I jumped out to fight the fire.

The scene on the front lawn was chaos. Piles of hose lay everywhere, an engine crew was rushing in the front door, and two drivers were frantically stretching a supply line. I saw Engine 16's captain, screaming orders at the top of his lungs. Suddenly, I was conscious of Spongy beside me, serenely taking it all in.

"Okay," he said, almost nonchalantly, "let's get a ground ladder and set it up under that window." Instantly, I felt myself relax. He was an island of calm in the midst of the panic. We slid a thirty-foot extension ladder out of the rack, and, stepping over hose, we soon had it leaning against the house under a second-floor window.

"Let's help Sam get his truck set up," he directed me. I hurried over to where the driver was working, grabbed the steel plates out of their holders, and set them on the ground on either side of the truck. Sam worked the levers, lowering the heavy hydraulic jacks into position on the waiting pads. Once the weight of the vehicle was fully supported, he climbed up to the turntable and began extending the ladder toward the burning roof. That done, I turned to look for my lieutenant. He was squinting at the house, where a window had just failed and black smoke and orange flames were spewing out.

"All done?" he asked cheerfully. "Okay, next assignment. We're supposed to go in and help the attack crew."

He and I masked up and stepped carefully over hose, up the front steps and into the darkened interior. I kept my hand firmly on top of his air-bottle, as he led the way into what appeared to be a living room. The crew from Rescue 16 were crouched at one end, directing a hose stream into a hole in the floor.

"Careful," one of them called as we approached, "it's burned up this way. We're not sure what's above us."

"Try opening a hole in the ceiling," Spongy directed me. "See if it's burning up there."

I took a stab with my axe. A shower of debris cascaded down on us, and the room went black.

"Out! Everybody out!" someone yelled. I collided with a body in the dark.

"Back out! Into the hallway!" The shouting continued. It was getting hot. With painful slowness, the traffic jam of firemen funneled out of the room. In the hallway we could see a little, and we regrouped.

"Ladder 16!" I called out, looking for my officer.

"Over here," I heard him reply. We found each other and stood behind the rescue crew as they doused the room from the doorway. Smoke began to clear out the broken picture window, revealing the charred remains of a couch, a coffee table, and a television, all covered with fallen drywall. At that moment, my low-air alarm went off, and Spongy gave the command to leave the building. At rehab, we switched our bottles, and a new assignment came over the radio. We were to re-enter the building and do a search of the basement.

We donned our masks again and entered the charred hallway. A narrow door under the stairs led to the basement, and we picked our way down carefully. A heap of burned clothing with a fallen lamp on top revealed the source of the fire, and the hole in the living room floor lay directly above. After a quick search of the basement, we found nobody. Returning to the main floor, we met the crew of Engine 15 coming down from the second floor. One of them was cradling a small object in his arms.

"Here!" he said, "take this outside." He passed the bundle to me, and I saw that it was a small dog, limp and smoke-stained. I scooped it up and hurried out into the sunlight. A chief caught sight of me and waved me over.

"Put it with the other pets." He ordered, pointing to a tarp that lay by the roadside. There were several lumps underneath, and I saw a few sets of paws sticking out. I lifted the tarp and saw two full-size canines, a lap-dog of sorts, and a cat, all lying stiffly in a row. I added the newest victim to their ranks and replaced the tarp.

"Ladder 16 from Command." The district chief's voice came over the radio.

"Go ahead," I heard Spongy reply.

"Get a saw up on that roof and make a vent hole!"

I ran quickly to the truck and opened the compartment. It was empty. Another crew must have grabbed it. I ran to Engine 16, but it too had been ransacked. A hurried search of two more trucks found only more empty compartments, and by now the chief was raising his voice.

"What's the delay? Ladder 16, get a saw on the roof—now!"

Just then, I spotted one. The RIT crew was standing on the front lawn, a tarp with their equipment lying at the ready. A chainsaw was among the tools. I snatched it up and made for the ladder. Spongy met me at the turn table, and the crew of Ladder 14 arrived at the same time.

"We'll take it," their lieutenant offered. Spongy nodded, and I handed over the saw. They clambered up, and soon I could hear the roar of the motor as it cut into the roof. Fresh smoke poured out.

"*It's getting into the attic!*" the chief was yelling. "*I need another crew to advance a hose line to the second floor!*"

It was two hours later. The fire was finally knocked down. The upper floor of the house lay open and gaping, the roof burned through, the walls half gone. I opened the cab of Ladder 16 and prepared to climb in. Suddenly, I halted. A glass box was sitting on the floor. Inside the box was a python.

"Hey, Lieutenant, look at this!" I called. He came over and peered in with interest.

"A snake!" he exclaimed. "Must be one of the pets. I guess another crew put it here for safe-keeping."

"Is it alive?" I asked.

"Give it a poke and see."

Gingerly, I prodded the reptile with an axe handle. It didn't stir.

"Dead," said Spongy, "I'm not surprised. They don't do well with smoke. Let's get it out and stash it with the other pets."

That done, we returned to the cab. Sam had lowered the ladder and stowed the jacks. We pulled away from the scene, and I had a last view of the funeral mound, the row of lifeless legs still sticking out pathetically.

"How did everything go outside, Sam?" the lieutenant asked.

"All right," he replied. "After you guys went inside, Rory

O'Connor came over and tried to give me a hard time. Something about a saw? Did you guys steal anything from the RIT team?"

"I know who that was," said Spongy, turning around to wink at me. "But I'm not telling anyone."

"Well, he went up onto the roof and started yelling at the Ladder 14 guys. Then he took it back."

"Who's Rory O'Connor?" I asked.

"He's the captain at Station 6," replied Sam. "They were assigned as RIT."

"Oops," I said. "I didn't know that was taboo. The chief was screaming for a saw, I grabbed the first one I came across."

"Yeah, we don't touch the RIT equipment," Sam said meaningfully. "But you know that now. Captain O'Connor is a real hot-head. He makes a big deal out of everything. They call him Roarin' Rory for a reason."

"You might get a cigar from the chief," Spongy teased. "Cigar" was a loaded term. In times past, whenever a firefighter made a serious blunder, he would be invited to the office, told to sit down, and be handed a cigar. The chief would then yell at him for as long as it took to smoke it. Since then, "getting a cigar" had become the ultimate threat.

"Well," I said. "There's a first time for everything. I've had a few firsts today. First time having a ceiling come down on me. And definitely my first time meeting a python on a fire truck."

11

STORM CLOUDS

My cell phone rang one Wednesday morning in March. I was on holiday, and I expected to hear the results of my transfer request when I returned to work. I was surprised to hear the voice of Chuck Brown, my affable lieutenant, on the other end.

"Hey, buddy, how are you?" he began. "Listen, I got some news. The District Chief stopped by this morning and said you're going to 6. Congratulations!"

I was speechless. Number 6! That was Roarin' Rory O'Connor's station. I certainly hadn't bargained for this. Station 6 was considered one of the best in the city, because of the number of fires it responded to annually, but I had other views on the subject. After the personnel issues at 17, I had carefully submitted for stations based on what I knew about the officers and crew. I had avoided 6 on purpose due to the captain's reputation. In typical fire department fashion, none of my requests had been approved.

"Oh," I said dully.

"Hey," Chuck went on, "don't worry about Rory. I know he's loud, but he's a nicer version of himself. He's been read the riot act, and he's not nearly so hard on the men now. You'll do fine down there."

"Thanks," I said, "I hope so."

"Yeah, it'll be fine, and you know what? You'll be having so much fun, you won't want to leave. You'll be getting tons of fires, lots of fires. I'm jealous," he laughed. "Now listen, give me a call in a couple of weeks, and let me know how it's going."

"I will, thanks, LT."

"Take care."

I hung up and exhaled. Out of the frying pan and into the fire, was the expression that came to mind. I had left one situation to get away from a bully, and here I was landed with a worse one.

Why, Lord? I prayed. For the last year I had been praying for a crew I could get along with and a captain I could trust. It seemed like a strange answer to prayer.

Then the realization dawned that this might be a test. Maybe I was being called to go through something I would not have chosen for myself. And with that realization came the conviction that I must do my best, however hard it might be. Here was an invitation to become a better firefighter, learn courage, and maybe grow a little through the process. I knew it would not be easy.

Station 6 was located across the river from the downtown core. The district was known for being the rough part of town. Rundown apartment complexes competed for space with derelict commercial buildings, while on the sidewalks could be seen many homeless people and drug addicts, a testament to the hard living conditions in the east end. Engine 6 was kept very busy, running to overdoses and the frequent structure fires that made it such a coveted spot for action-hungry firemen. Captain O'Connor had grown up in this neighborhood and had inherited its hard edges.

My first morning at Station 6, I showed up a little early and began checking the equipment on the engine. I heard the door open at the back of the truck bay and a heavy tread coming across the concrete. It was my new captain. Rory was well over six feet four, with iron grey curls and a large

Roman nose. But it was the eyes that caught my attention. Large, intense, and penetrating, they emanated a presence that was both powerful and volatile. He shambled toward me with a bit of a stoop, looking out from under his eyebrows like a bear disturbed while hibernating.

"Captain O'Connor?" I said loudly, feigning a confidence I did not feel. I shook his hand firmly.

"Hi," was all he said, but a grin showed for a moment on his weathered face.

We met again in the kitchen after morning truck checks.

"So," he rumbled, "I hear you're religious."

Here we go, I thought. "Yes," I replied briefly.

"I grew up Catholic," he said, a look of reminiscence spreading over his face. "I don't go to church any more, but I consider myself a spiritual person."

"Mm, hm," I murmured.

"Any kids?" he asked. He seemed to be trying to put me at ease.

"Yes," I said, warming. "Three girls."

"Three girls," he grinned. "You're a busy man. I have two kids myself, but they're grown up now. One thing kids need is lots of love. No one can live without love. You seem like the type who'll raise them well."

I was a little taken aback by this thoughtful foray, which seemed at odds with his reputation.

"If anything bothers you here," he finished. "Let me know."

"Thanks," I said, still surprised. "I will."

The first shift passed without much incident. The captain laid out his expectations for his nozzle man and tool man, and watched approvingly as I practiced deploying the 200-foot attack line. The only hint of his hard side came out when we were discussing tactics.

"And make sure," he said forcefully, "that you choose the

right line. If the fire's on the third floor, there's a chance we may need the 400-footer. It's a pain in the ass to move and takes up time. If I see it deployed, there better have been a good reason for it."

I sensed the menace behind these words. However, nothing of the sort came to pass that day. We spent the remainder of the shift responding to insignificant calls: a fender-bender, alarms at the hospital, wires down. I drove home after shift change with a feeling that life at Station 6 might be bearable after all.

It didn't take me long to experience one of his famous outbursts. It was my second shift, and we were driving around the district on a routine familiarization run. All of a sudden, the radio cut in.

"Engine 6, I need you for a structure fire in Station 3 District. Address is 675 Lafrance."

I pulled on the straps of my breathing apparatus and reached for my helmet. The truck lurched as we swerved out into traffic with the siren on. The captain was commanding the driver in a loud voice.

"Take the highway! Go around that car!"

Buckles clanked as my partner Eddie fastened himself into the straps of his SCBA and checked his facepiece. He reached out a fist, and I gave his a tap with mine.

"All right, let's do this!" he said cheerfully. I couldn't muster up the same enthusiasm. I felt my stomach tightening into a knot. My first fire with Roarin' Rory. Could I prove myself without calling down the famous wrath?

"Get off here!" the captain was bellowing, and I felt the truck swerve off the freeway onto an off-ramp. Radio chatter from the fire was filling the cab.

"Engine 3 and Ladder 3 on scene. You can confirm a working fire. We have heavy black smoke visible on side two, open flames issuing from a basement window."

"Copy, Ladder 3, Car 1 almost on location. I'll establish command. Engine 1, I'll get you to catch the hydrant on the corner of Lafrance and Summer View. I have Engine 3 making entry on side 1, they'll need water."

We hurtled past the end of the ramp, and I jerked back heavily as the driver hit the brakes. We began weaving through congested side streets. From my rear-facing seat, I had impressions of tenements flitting past and vehicles edging up onto the sidewalk to let us by. Rory's window was down, and he was hurling profanities at any drivers who were slow to get out of the way. One final turn, and we were in view of the fire.

"Oh, yeah!" yelled Rory. "It's a good one! Look at all the black smoke!"

The truck halted. We were about a block away. Rory turned to us.

"Ben, Eddie, grab a chain saw and some forcible entry tools. We're going to the front door."

I was out of the cab with a bound. It only took a few seconds to collect the tools, but the captain was already halfway down the block at a run, and I sprinted to catch up, the tools banging awkwardly against my legs. He halted in front of a two-story, red brick duplex. Smoke was pouring out every window, and firefighters from Engine 3 were hauling a hose line through the front door.

"Get your mask on!" he bellowed at me. I dropped my tools, donned the facepiece, and felt a rush of cool air as I switched on my bottle.

"Force that door!" he barked. I saw a second entranceway, belonging to the other half of the duplex, with a barrel lock and deadbolt. I seized the Halligan and inserted the bill into the crack between the door and frame. I pushed down hard, but it seemed unwilling to yield. Eddie was beside me, giving direction into my ear.

"Closer to the lock. There you go. Together now." With our combined strength, we leveraged the tool, and with a cracking, splintering sound, the door gave way. Smoke swirled out to greet us.

"Get in there," roared the captain. "Do a search!"

Smoke was banked down to the floor, so I hopped onto my hands and knees, as I was trained to do, and crawled into the dark. I bumped into something hard.

"Stairs here! Stairs here!" I called out.

"Get up them!" came the captain's answering shout. "Search the second floor!"

I crawled upward sounding each stair with my tool to check if it was solid. Memories of training exercises came back, and stories of firefighters who had fallen through partially burned out stairs by forgetting to check as they climbed. But the wood gave a dull solid noise, and my tool sprang back with each blow. Reaching the landing, I groped forward and found a door. It was locked. A brief effort with the Halligan, and it swung inward, revealing a small apartment, empty. Eddie had found the other unit, and we almost bumped into each other in the hallway.

"All clear upstairs!" he called down. O'Connor's voice roared at us from below, ordering us to come back. We clattered down stairs and out into the daylight.

"Into the fire unit!" was the next command. "We're looking for the basement stairs."

It was even blacker in there, if that is possible, and my ears were full of the noise of chainsaws and the shouts of the attack crew, as we jostled our way down the narrow hallway. We halted in some kind of room at the back of the house. I sensed the captain brushing by me in the dark, going back the way we came, but nothing was said. I turned to follow, but at that moment my flashlight beam shone for a second on what looked like a door handle. I reached out and felt it. Yes, it was a door.

"Basement stairs!" I called out. I wedged my tool in the crack, ready to force the door. Eddie heard me, and was by my elbow, but there was no answer from O'Connor. A faint inner warning bell sounded inside of me. Where was the captain? And that was where I made my mistake. Instead of following up and telling Eddie we needed to find him immediately, I decided I should force the door. It took only a minute. Just as the door swung open, I became conscious of a mask face-to-face with my own, and two intense eyes looming through the smoke.

"Where were you?" The full force of Captain O'Connor's personality was behind the words.

"I found the basement, sir." I replied breathlessly.

"You're not leading!" he roared. "I'm leading! Now follow me!"

I jumped to it, hurrying after him as he moved quickly down the black hallway and up the stairs. Eddie was close behind me. The captain pushed open a door, and we stepped into a small apartment, bare except for a bed frame with no mattress. The smoke was relatively thin, and we could look around us.

"Open up that floor!" O'Connor ordered. "Eddie! Where's the chainsaw?"

Eddie went flying back down the stairs and was back with the saw in less than a minute. The motor roared, and wood chips flew as it ripped into the floorboards. I worked behind Eddie with the axe, tearing up the broken chunks and pushing down the ceiling plaster to reveal the room below.

Focus, I kept saying to myself, trying to shake the mental image of those angry eyes. The captain's rebuke was still ringing in my ears.

You still have a job to do, I reminded myself, but I was becoming conscious of a mounting fatigue. My breath was starting to become labored, and with each stroke of the axe my arms felt heavier.

Don't give up, I thought. *Don't let him see you're tired.* All at once my low air warning bell went off. My bottle was almost empty. I felt a heavy hand on my shoulder.

"Go change your bottle!" came the captain's voice in my ear. "Then come back!"

I stumbled downstairs, glad to be away from him. At rehab, I switched my bottle, then reluctantly made my way back inside. They were ripping apart ceilings, and I set to, with my arms aching. At first, lath, plaster, and insulation rained down in a steady shower, but with each pull I felt my strength waning. I could see the captain standing at the door, appraising me. At last he growled:

"That's enough. Let's get out of here."

We descended the stairs one last time. The fire was out, and crews were moving throughout the building, checking for hot spots. After a brief consult with the chief, we were released from the scene.

It was a long ride back to the station. The crew engaged in cheerful banter, but I could only feel a gripping sense of failure. Freelancing was a capital sin on the fireground, and losing one's crew in the hot zone was something no one did, ever. I knew I was in disgrace. The captain took the first opportunity to take me aside and remind me in menacing tones that I had screwed up. I apologized, and we parted awkwardly.

I cleaned my gear, showered, and repaired to the kitchen for the evening meal. The delicious aroma of chicken breasts stuffed with cream cheese and bacon bits filled the station. Eddie had served up heaps of mashed potatoes and a sumptuous Greek salad to go with it, but I couldn't taste the food. For the first time in my career, I felt mentally and physically defeated by a fire.

There were no calls during the night, and I tossed and turned on my cot, replaying the events of the day. I rose early to make the coffee, and the captain found me at the kitchen table gazing into my coffee cup.

"So," he rumbled. "You got your first fire here already."

I nodded without much enthusiasm.

"Wish it had gone better," I offered. He laughed.

"Aw, that wasn't such a big screw-up in the big scheme of things. Eddie was there with you, and it's hard to see in there."

Suddenly, it seemed like I had gone up one end of the roller coaster only to slide down another. Was this man mentally stable? Yesterday, my mistake had been the end of the world, today it was no big deal. It began to dawn on me that he must have a very different emotional barometer than the average human being. Explosive, yes, but he was no grudge-bearer. The storm clouds gathered, the storm burst, then all was sunshine again. Already he was off on a new topic.

"I can't believe we're getting more snow," he roared. "Do you still have snow at your place?"

I drove home that morning with mixed feelings. I would have to learn to shrug off these outbursts, I thought, and keep better focus on the fireground. Still, it would not be easy going to work. This felt far crazier than anything that had happened at Station 17. I did have one consolation, though, and that was Eddie Baker. In addition to being thoroughly competent, my co-worker was also kind. I would arrive at work in the morning keyed up for the day, my stomach in a knot, and it was always heartening to hear him whistling as he gathered up his gear to set up on the truck.

"Hey there!" he would greet me, always grinning, always cheerful. "How's it going?"

Despite being six years senior to me, he always addressed me as an equal. We would chat as we went about our morning routine. Eddie was a former plumber and trucker and had an inexhaustible store of how-to knowledge, besides being an excellent station cook. He would inevitably wind up each morning conversation by stating: "It's gonna be a good day," in his broad country accent. I never felt any day could be good with Rory around, and it puzzled me how Ed could be so nonchalant.

I took the soonest opportunity to pick his brain about the captain.

"You've worked with Rory for a long time," I ventured. "How do you deal with it?"

He looked thoughtful. "Well," he said at length, "I guess I just decided that once I left the station in the morning I wasn't going to think about work anymore. Whatever happens here doesn't really matter after that. I can't control what Rory says, so why worry about it?"

It was a simple philosophy, and I envied him.

"I'll have to learn how to do that," I said. "I'm taking everything home."

The next shift I was scheduled to work was Easter Sunday. My wife and I took the children to the Good Friday service, and I was struck by the account of Christ's Passion. It made my troubles seem small, and I left the church with the conviction that I should bear my mercurial captain with good grace. The Easter Vigil came in a burst of glory, and the promise that after every cross comes the Resurrection. As I drove into the city Sunday morning, I mulled it all over, wondering how it applied to what I was going through.

At first things were quiet at the station. Rory set me to work doing small maintenance jobs, after telling the senior firefighters to relax.

"We got a new probie to do everything now," he said with a loud bark of a laugh, ignoring the fact that I was the same rank as they were, Firefighter First Class, and three years off probation. Apparently, seniority was all that counted at Station 6. Around 2 o'clock in the afternoon, the doorbell rang. I ran to answer and was pleasantly surprised to see Kate and the three girls on the doorstep in their Easter dresses.

"Come in!" I said, delighted to see them. I thought of Rory and wondered how he would treat my family. I certainly wasn't going to let him be rude. They stepped into the entryway, holding homemade Easter cards and looking around them curiously. At that moment, the captain came around the corner and stopped short. Instantly, his stony demeanor changed into what I took for an attempt at an expression of congenial hospitality.

"Hello, ladies!" he boomed pleasantly. "Thank you for coming! What's this? A card? For *me*? Why thank you!" He bent down admiringly while Beth held up her card. Kathleen and little Marie gazed at him with wide eyes, uncertain what to make of this red-faced giant.

"Hello," my wife greeted him, seemingly not intimidated. "I'm Kate."

"Hello, Kate," he replied cordially. "Kate, that's my daughter's name."

"Oh really?" she said pleasantly.

He accompanied us around the station, making friendly small talk while I gave a tour of the kitchen, the truck bay, and Engine 6. The visit was cut short by the alarm tones, and we were off on a smoke visible call. The girls watched with interest as we jumped into our bunker gear, and I had a last view of them waving at me as we roared out of the station and sped up the street.

The truck turned the corner onto Portage Avenue, and we could see a black column drifting into the sky. Fire trucks already lined the street, and from the radio traffic we knew that suppression activities were well under way.

"Portage Command, Engine 6 on location. What's our assignment?" asked Rory into his mic.

"Engine 6, I'll have you get a ladder to the garage roof and open up a vent hole over the fire."

When we arrived on scene, Eddie and I ran around to the back of the truck and slid the heavy steel ladder out of its rack. One on each end, we set off down the street at a quick trot. When we reached the front of the house, I could see a beehive of activity taking place around the garage. Smoke was still belching out the door, and the crew of Engine 15 was exiting with their low-air alarms ringing. Piles of charged hose lay everywhere, and the buzz of chainsaws sounded from the roof. I could see Rory up on top of the garage already, mask-less in the smoke, yelling at the crew of Ladder 16, who were trying to open up the fascia. We placed our ladder on the garage and clambered up with chainsaw and axe.

"Open up that roof!" Rory ordered. "And don't cut through the rafters!"

I fired up the saw and watched sparks fly as the chain bit into a roofing nail, then I was off in a straight line, cutting through the shingles and plywood. At first, smoke trickled up through the crack I had made, then it surged up in a mighty billow as the roofing fell away. I had opened up an area roughly 4 feet by 4 feet, and the ladder crew advanced to the hole with a charged hose line, ready to direct it into the garage below. The vent we had made would allow the smoke to lift, giving the attack crew better visibility, but we must be careful not to blow smoke and embers back down on them with our stream of water. The captain radioed to the attack crew, asking for clearance to direct a stream downward.

"Go ahead," came the reply. "We've exited the garage."

"Give it a shot!" the captain ordered. Soon a blast of water was cooling the timbers beneath us. We took turns with the nozzle, switching out whenever a man got tired, and before long the smoke had cleared away, leaving only charred rafters and trailing wisps of steam. Shutting off the nozzle, we surveyed the damage. The blackened bulk of a car lay in the wreckage, half-buried by shattered drywall and half-burned bats of insulation. Firefighters from Engine 15 were beginning to wade into the mess, dousing hot spots and pulling apart smoldering debris as they went.

At that moment, I spotted a white helmet directly underneath. The District Chief, without his mask, his coat hanging open, frowned up at us.

"Don't look!" he barked, approaching what appeared to be the remains of a bathroom. The toilet was partially buried, but he kicked aside the burned boards aggressively.

"And don't soak me, either!" he warned.

After a short cleanup, we were on our way back to the station. My family had left, but I looked forward to telling them all about the fire when I got home. After we checked our tools and SCBAs, Eddie headed to the kitchen to start working on supper. Already my stomach was grumbling, and I followed to help speed up the process.

"Chicken parm tonight," he announced, filling a huge frying pan with oil. He set me to work pounding four large boneless, skinless chicken breasts flat and dipping them in egg-yolk. Each piece he then rolled carefully in a mix of breadcrumbs, salt, and parmesan cheese and dropped it into the hot oil to deep fry. He opened the fridge and lifted out a giant pot of spaghetti sauce, his creation from the shift before. Soon it was bubbling away merrily, filling the kitchen with a rich aroma that made me ravenous. Eddie's sauces were legendary.

"It's gonna be good!" Eddie remarked, grinning.

Once the chicken breasts were golden brown, Eddie lifted each one out carefully, poured a ladle-full of the sauce over it, and sprinkled a generous pile of melted cheese on top. Then the breasts were put on a baking sheet and set in the oven for the cheese to melt. Meanwhile, I was watching the noodles, ready to drain them as soon as they were *al dente*. At last everything was ready, and we each sat down to a mountain of spaghetti topped with the thick, rich sauce, alongside a hearty slab of chicken covered with more sauce and melted cheese. Every station meal comes with the uneasy knowledge that the tones could go off at any moment, and one never knew if there would be another chance to eat that day. But today we were able to finish our meal in peace. Hunger is the best spice, and my appetite was given an extra boost by the knowledge that I had done well at the fire. Rory's parting comment as we drove away from the fire scene had been: "Good job, men."

I was destined to go into the hot zone one more time with Rory. It was the very next shift, and this time when the tones went off, I buckled myself into my SCBA with a greater confidence. Eddie gave me the habitual fist bump, and we were off to yet another fire.

"Where are we going?" Eddie yelled above the siren and air horn.

"735 Pallisade," the driver yelled back. "5's district. We'll be second-in."

Second-in meant that our driver's job would be to establish water supply for the first-in truck, and we would be heading to the fire itself to assist the first-arriving crew. Smoke was visible a block away, and we came in view of a three-story row house with flames through the roof. Engine 5 was parked out front, and its driver was hurriedly pulling off a supply line. A crowd of spectators had already gathered on the street. Some were jumping in to help move the heavy hose, others stood there with their cell phones out, filming the excitement.

"Help him with the hydrant," Rory ordered our driver. "You two come with me."

We ran up the front steps and halted in front of a set of doors. Before I could use my tool, the captain turned around and with a furious kick knocked the door open. A second kick vanquished the other one, and without waiting to don his mask, Rory ran inside. I followed, hurriedly groping for my own facepiece. The air was clear in the foyer, but up above I could see a kind of indoor balcony, where a thick layer of smoke was beginning to accumulate. Rory paused at the top of the stairs, heedless of the smoke.

"Bring a line!" he ordered. Eddie turned and ran for Engine 5, and I followed. It took us a moment to bring up the 400 feet of attack line, and Rory was masked and ready when we arrived. I clattered up the stairs, clutching the nozzle in my gloved hands, while Eddie stayed at the door, pushing the hose in after me. I looked around. Where was the crew from Engine 5? This was the fire unit, and they had arrived on scene first. They must have gone through a wrong door. I followed the captain up to the first landing. Smoke swirled around my head. Voices sounded on the stairs below. Engine 5's crew had arrived just a moment late, after entering the neighboring unit and finding no fire there. They were dragging their own hose line up the stairs behind us. I followed as Rory led the way up another flight of stairs, and we were on the third floor.

"Stop there!" he ordered.

I took a breath and looked around me. We were on the balcony, with two bedroom doors in front of us. Black smoke was banked down almost to the floor, but from where I crouched, I could see a bed in the nearest of the rooms, orange flames shooting from the mattress. It was strangely quiet. The only sounds were the faint crackle of flames and the muffled footsteps of the engine company a floor below. The uncharged hose lay slack in my hands. Suddenly, the smoke dropped, and I lost visibility. I felt a blast of heat.

"Get out!" the captain roared. "Back down the stairs!"

I scuttled backward as fast as I could and crashed into someone. Rory was bellowing on the stairs as the heat intensified.

"Get out of the way! Move!"

Who was he yelling at? I heard other voices and felt movement behind me. We had collided with Engine 5 in the dark, and there was a traffic jam halfway down the stairs.

"Charge the line!" He began to scream. I don't know if he was using his radio or not, but it wouldn't have surprised me if they had heard him out on the street. I felt the line stiffen.

"It's charged!" I called over the din. It took a few tries before he heard me.

"Advance!" he roared. "Up the stairs! Spray the ceiling! Cool it down!"

I knew the tactic. Smoke piling up inside a building creates a thermal layer, with the hottest smoke near the ceiling. Eventually it can become so dense and hot that the smoke itself will catch fire. If the room gets hot enough, everything in that space will spontaneously ignite, creating the dreaded event known as "flashover", which has killed many firefighters, despite our sophisticated gear. Directing small bursts of water at the ceiling cools the smoke and delays the onset of flashover. Dragging the hose back up the steps, then, I sprayed several short bursts upward. We reached the landing, and I began a creeping progress toward the source of the heat. The fire had increased in volume, and a dull orange glow illuminated the space around us, despite the thick smoke. I heard a scuffle behind me. Whoever was on the other nozzle seemed to the think the command to advance was for him, for Rory's voice was bellowing through his mask.

"Not you! Get away from me! Move!" In the flickering half-light, I saw his arms flail out, and the firefighter, still clutching the nozzle, went tumbling over in a heap. Next moment, the captain was beside me, barking orders at my head.

"Into the bedroom! Hit that bed! Hit the ceiling!"

As I obeyed, the fire died with a hiss. The room darkened.

"Shoot your stream out the window! Ventilate!" I aimed the nozzle at the open window, and my jet of water streamed through it, drawing the smoke after it. It was another tactic I had learned in drill school. There were firefighters surrounding me now, pulling down drywall and exposing pockets of burning insulation. Engine 5's nozzle-man, apparently recovered, was dousing hot spots. It appeared the fire was out.

All at once, there was a noise like an express train, and a shower of drywall and insulation rained down, followed by a terrific deluge of water that beat against my helmet like a sledge hammer. I staggered back against the wall, clutching desperately at the hose line, while what looked like Niagara falls poured through the broken ceiling, filling the room with flying debris. I could hear my captain shouting into his radio:

"Shut down the master stream! We're getting hammered in here! Tell Ladder 5 to shut down their gun! *Shut it down!*"

Then I knew what was happening. Ladder 5 had raised its aerial device into position above the fire and had opened up with their large-caliber nozzle, letting loose 1,000 gallons per minute. There was no need for my hose line now, the room was swamped. Rory's transmission must have gotten through, however, for in seconds the torrent slowed to a trickle, and I was left standing in a ruined room, almost up to my ankles in water.

"Get out of there," the captain barked. "We're done."

I backed out of the room, soaked to the skin. It was a slow slog down the stairs. We were walking down a gentle waterfall, and hoses still lay on the stairway. We picked our way down, and I blinked as we emerged into the April sunshine. The street was clogged with fire trucks and personnel, but Rory had already picked out his next victim in the crowd. I watched as he advanced on Ladder 5's lieutenant. The officer looked up and hastened to forestall the attack.

"I know! I know!" he called out. I didn't catch Rory's reproach, but I could see him waving his hands around emphatically. Luckily they parted without coming to blows.

We switched our bottles, and as we were gathering up our tools, Engine 5's crew came over to join us. Rory broke into a toothy smile.

"Was that you I pushed on the stairs?" he guffawed. "You must've thought I was telling you to go up! Haw haw!"

He slapped the nozzle-man hard on the back, and the young firefighter grinned sheepishly in return. Then turning to us, Rory gave his last order of the day:

"Help roll some hose, guys. Then we're outta here!"

Rolling hose didn't take us long, and half an hour later we were driving back to the station. My limbs felt heavy with fatigue, but there was a sense of accomplishment stealing over me. I had worked three fires with Rory now, and with each one I was becoming more proficient. I felt I now had a sense of his expectations, and the wild outbursts seemed less intimidating.

"People get so worked up at fires," Rory said loudly as we drove along. "That's always been my thing, helping everybody calm down."

It was evening, supper was over, cleanup was done, and I was in the truck bay watching the sunset through the bay door windows, when Rory came shambling over to me.

"I'd like to move you to the driver's position next shift," he said.

"Yes sir," I replied automatically, but I felt my heart sink. Fighting fires with him was one thing, but driving the fire truck was quite another. I had seen him in action enough to know what I was in for. He was loud even toward experienced drivers, and I could expect nothing short of abuse throughout the learning curve. Emergency response driving comes with its own level of stress. A driver must know his route, be fast, and be safe while driving a large vehicle through traffic. Then at the call site, the pressure to supply the crew with water in a timely manner falls to him. I had already felt the weight of this responsibility while at 17, and the prospect of having a hothead in the seat beside me was anything but pleasant.

"We'll practice catching hydrants next shift," he finished. "To get you up to speed."

The next morning, a silver Sprinter van pulled into the station parking lot. The door slid open, and a bearded man of about sixty stepped out.

"Hello, son," he greeted me.

"Hi, Dad," I replied. We hugged, and he stepped back to look me over.

"You ready for the pilgrimage?"

"I sure am," I said. "I need it."

"How was your shift?"

"Pretty crazy. We had a big fire."

"Well, maybe you can nap in the van."

I climbed in and was greeted by ten other men, ranging in age from sixteen to seventy.

"Hello, everyone," I said. Many of them I already knew, and there were a few introductions and handshakes from the new faces. Soon we were cruising down the highway, and I felt my head nodding forward. Images from the fire swirled through my mind. We left the city behind and were traveling through open country, the bare fields lying brown under a gloomy spring sky. I must have fallen asleep, for when I awoke, we were coming into another built-up area. Commercial plazas and warehouses sprawled in all directions. Graffiti-covered overpasses towered above us, their chipped and crumbled edges bearing witness to the age of the tired city. After another fifteen minutes, rows of high-rises appeared, mounting the slope of a small mountain, at the very top of which loomed the dome of an enormous church. We took an off-ramp and were soon traversing narrow streets, each turn taking us higher and higher. A last turn into a parking lot, and we were coming to a halt under the shadow of the basilica. The door slid open, and pilgrims began climbing out, yawning and stretching their cramped limbs.

"Who brought knee pads?" someone asked. There were some chuckles.

"You don't bring knee pads on a pilgrimage," someone else rejoined. There was a general movement toward the church. At the edge of the parking lot, the group halted at the foot of some cement stairs.

"All right, let's begin," said the leader. He knelt down on the first step. Looking up, I saw that the stairs stretched up the steep hillside all the way to the doors of the church. A step had been built for every Hail Mary of a twenty-decade rosary. Our annual tradition was for fathers and sons to come here on the feast of St. Joseph the Worker and pray our way up the hill on our knees. As we climbed, one hard stair at a time, I kept seeing the face of Rory O'Connor in my mind.

I'm carrying him up the hill, I prayed silently. *And asking you for help.*

That evening, driving home, I overheard one of the dads discussing an upcoming camping trip.

"I'm taking my sons," he said. "It's cold this time of year. No extra frills though. Boys need hardship to become men."

His words struck me. *Boys need hardship to become men,* I repeated to myself. Maybe that's what being at Station 6 is about.

I spent the next few days mentally preparing for the driver position, reading maps and telling myself that I would be fine. Sunday night rolled around, and I packed my kit bag in preparation for the following day. It was then that I noticed an uncomfortable feeling starting in my chest. It began as a warm glow, but as the evening progressed it became sharper and spread to my shoulders.

"Stress," I said to myself. "How foolish. You just need to go for a run."

I set out down the dirt road and tried to distract myself by looking at the trees, the sunset, the hay fields lying golden in the dusk. I could not shake the feeling of impending doom. I tried to tell myself that what I was experiencing was disproportionate to the actual threat.

"What's the worst that could happen?" I muttered. "Getting yelled at? Not a big deal."

In bed, I dozed off and on and woke before my alarm. Getting up, I went to the window and looked out. Mist covered the pasture, and the birds were beginning their morning chorus. It seemed so peaceful, so at odds with the turmoil in my mind and the ordeal that was waiting for me in the concrete jungle. I rubbed my chest. There was no doubt about it, it still felt tight and hot.

"Everything okay?" my wife asked in a sleepy voice. I hesitated.

"Bit of chest pain, I think," I said.

She sat bolt upright. "Oh, honey! You need to go to emergency right away."

"It's nothing," I replied. "Just a bit of stress. Rory wants me to start driving today; I'm just a little nervous about it, that's all."

"I think you should get checked out," she said, a worried frown on her face.

I sighed. "If I book off sick that's just running away. I'll have to face it next shift anyway."

"You have your health to look after," she said firmly. "Please, for your family's sake."

I knew she was right. I should never have allowed stress to get to this point. Accordingly, twenty minutes later I was pulling up in front of the local hospital. A young resident checked my blood pressure, as I briefly explained my troubles.

"That's no fun," he said sympathetically. "We'll run some tests though, just to be sure there's nothing wrong with your heart."

After an X-ray and a cardiogram, I was sent home with a recommendation to read self-help books and manage my stress. I knew there was nothing for it but to face my fears and return to work. Two days later, I dropped my boots beside Engine 6 and checked the duty board. I was not assigned to the driver's position. The captain came in, glanced at me briefly, and began his morning checks. I walked around the truck, absently opening compartments and lifting out tools, feeling the mounting tension that always accompanied a day at Station 6. I turned to see the captain looking me over.

"You were off sick Monday," he began. "Everything all right?"

"Chest pain," I said. "Nothing serious."

"Hm, that sounds pretty serious. Did they do any tests?"

"Yes," I replied. "But they didn't find anything. I was told to manage my stress, that's all."

"Stress, eh? Something going on at home?"

I realized I would have to come clean.

"No," I said, looking him full in the face. "It's work. It's this station, the driving, everything . . . " My voice trailed off. I couldn't quite bring myself to say, "It's you."

He was silent for a moment, but I could see that he knew what I meant.

"I'm not here to bust your balls," he said at length. "I'm here to help you along. You're doing well, but I don't want you driving if you're not ready. I'm not going to coddle you, you know me well enough by now to know that," he laughed shortly, "but I'll help you succeed. Just let me know when you're ready."

He turned on his heel and walked away. I watched him, absorbing this new turn of events. At least now we understood each other. Perhaps this would be a turning point, and he would tone down the aggression from now on. There was more to come on the subject, however. That afternoon he called me into the office.

"Shut the door," he said. "Have a seat."

I complied, not sure what he was going to do.

"Stress gets into your mind. It gets into your soul," he began. I nodded. "You're very new on the job. Four years is very early to be having these issues. It's a tough job," he went on, "and it can mess with your head. I think you need to get this dealt with now, before it becomes a bigger issue in your career. I look on that as my responsibility, as your captain."

I nodded again. There seemed to be nothing I could say.

"Now," he said, "I have a list of phone numbers for you. It's our mental health support team. 100% confidential, and they're all firefighters, they know what it's like."

He handed me a sheet of paper.

"There's no shame in asking for help," he said. "I've had to do it."

"Thank you," I said simply.

"And just so you know," he added. "I have no issue with your job performance. The guys seem happy to have you in the station. You have some strong points. You got some weak points, too. You're a hard worker, I can see that. If there's an issue, we'll talk about it and move on. So . . . deep breath, everything's good."

I didn't see Rory for almost two months after that. I went off on my spring holidays, and when I returned to work, he was off for much of the summer. We had a variety of acting captains filling in while he was away, and with less volatile personalities in charge of Engine 6, I felt confident beginning my tour as driver/operator. The fateful fire at 281 O'Donnell occurred during this time, which I looked on as a coming-of-age point in my career. Following Rory's advice, I also began regular phone conversations with a member of the mental health support team. Slowly, the realization was dawning that taking care of my health, both mental and physical, and consequently my family's well-being, was as essential an ingredient to becoming a successful firefighter as training and experience. It was in a much more stable state of mind, therefore, that I walked into the station for my first day back with Roarin' Rory. I had been detailed to another station the shift before, which incidentally had been the captain's first shift back at Number 6 since his holidays, so we had not crossed paths.

I said good morning to Eddie as I hung my coat on the driver's door and placed my helmet on the dash. There was no sign of the captain yet.

"Well," said Eddie. "You drove Rory to an early retirement."

"What?" I blinked, uncomprehending.

"Too bad you weren't here for it, but Wednesday was his last shift on the job."

I looked at him, stunned.

"You mean he's gone?"

"Yeah! He never said anything to us about it all day, but then just before shift change he came into the kitchen, shook hands, and said, 'this was my last day on the job.'"

"Really!" I said.

"Yup! He's using up the last of his sick days, then it's official retirement."

It took a few moments for this to sink in. I stared out the windshield, watching my future change before my eyes. So, all that worry had been for nothing. What could I say, or think? The ordeal was over.

I thought a lot about Rory over the next several weeks. What a strange combination of hardness and understanding, anger and compassion. He was a bully, there was no denying it, and I had been witness to the regular outbursts of his passionate personality. Simply being in proximity to him had brought out the worst anxiety I had ever known.

Overcoming that anxiety had been an important step for me, but I still harbored some resentment toward the man who had caused it. What made a person like that, I wondered?

I had one last meeting with him, about a month after his announcement. He had decided to drop in at the station, and we were all gathered around the kitchen table, listening to his retirement plans. The conversation drifted naturally enough to bygone days, and he began to regale us with tales of thirty years ago, when men fought fires without masks, with only a tin helmet, raincoat, and rubber boots to protect them. He reminisced about riding the tailboard of the firetruck, the camaraderie, the pranks, the friends lost to cancer, the burn victims, the steady toll that years of witnessing trauma took on the mind.

"What a job!" he finished. "What a job!"

"Will you come back for a shift or two?" asked Eddie.

"No," he replied emphatically. "I'll never come back. When you feel the tears coming to your eyes on medical calls, you know it's time to quit. The hard drive is full."

There was silence. The subject changed to lighter things, but I sat there thinking about what he had just said. He had fought hard all his life, and in the end, no amount of fighting could spare him from what he had seen. For the first time, I felt compassion for him. I had suffered a little, but my experiences were nothing compared to the load he carried. And with that knowledge came the ability to forgive the man who had controlled me with anger, caused me to doubt myself and experience fear. I knew I would never be like him, but I could understand him, for I had carried for a short while a burden like the one he had carried his whole life. His had made him into something hard. I knew I would have to make a different choice.

12

THE BREAKING POINT

I thought that my symptoms would go away with Rory's retirement, but it was not to be. Three shifts went by, and I still felt the tightness in my chest as I drove in to work. Part of the reason was Bryce Halton. Bryce was our senior firefighter at Station 6 and had spent the majority of his career on Rescue 2, a very busy truck in an action-packed district. He was one of the most competent and knowledgeable people I had ever met, but if he had a fault it was an unspoken disdain for my generation.

"These new firefighters just don't seem to get it," I overheard him remark to Rory one day. "Nobody has any work ethic anymore. Everybody's a victim."

There may have been truth to the observation, and I tried my best to make a good showing by working hard. However, as the weeks wore on, I was losing my desire to please him. His manner was simply making it too difficult. Unlike Rory, he was not hot-tempered. He was generally cool, approachable, and well-spoken, but his attempts to give me advice and direction had become increasingly condescending, even supercilious. At first I barely noticed this dynamic, being far too consumed with Captain O'Connor. However, once Rory retired, Bryce stepped up into a greater leadership role, and his high-handed demeanor came into sharp focus.

One morning we got a call just as I was conducting my

morning truck checks. I was the driver that day, and as always when the tones went off during morning routine, the checklist would have to wait.

We pulled up on a vehicle fully engulfed in flames, and I switched the transmission into pump gear, ran to the control panel, and charged the line just as Eddie and Bryce were donning their masks. Our officer of the day, Captain Mike McCannell, stood back to assess the scene.

"More hose!" he called. Usually the attack crew would deploy enough line to make a full circle around the vehicle, but this morning they pulled up short. I hurried around to our truck's front bumper, where the rubber-jacketed hose reserved for car fires was stored. There were no extra lengths in the compartment.

"That's it!" I called back. He waved back and gave the order for the others to start the attack. They hosed down the car, cooling the gas tank from underneath and dousing the engine compartment from a few feet away. I could feel their frustration. Not having enough hose was hampering their efforts. It took some time, but eventually the fire was knocked down. I helped them drain the hoses and repack the front bumper.

"Why are we short?" Bryce asked cuttingly.

"I'm not sure," I answered. "There are only two lengths in there for some reason."

He didn't reply. I remembered checking the front bumper that morning, but perhaps not thoroughly enough, for I hadn't taken note of how many sections of hose were folded in the compartment.

We climbed back into the cab, in time for another call to come over the radio. We were needed for an MVC on the other side of the district. It was a minor fender-bender, and the drivers seemed uninjured. Bryce checked them over while Eddie disconnected the batteries and helped the tow-truck driver get hooked up. It wasn't long before we had cleared the scene and were returning to station. I was just backing the truck up to the bay doors when a third call came in. Away we went, this time heading to the hospital for alarms. It didn't take long to confirm a false alarm, reset the panel, and head home. By now the morning was half gone, and I was looking forward to finishing my truck check. I backed into the station, set the parking brake, and shut off the engine. The others began to clean their gear, as I retrieved a spare length of hose from the storage rack and repacked the front bumper. Bryce came over to join me, and we arranged the last section together. When we were finished, he turned to me.

"I assume you're like me," he began. "You don't like nasty surprises. That was a nasty surprise this morning. Things like that don't happen when you check your truck properly."

I opened my mouth to reply, but he kept going.

"Also, have the saws been started yet?"

"No, not yet," I replied.

"It's ten o'clock in the morning," he said severely. "We're at a busy station, the saws haven't been started, and you haven't checked your truck. There is no excuse for failing to do your job. Maybe you've gotten used to doing things a certain way at your old station, but here everything has to be done properly, first thing in the morning, and no matter what. If Rory were here, he would have lost it over what happened. And he's right. I know I'm losing patience with people not doing their job."

I felt a surge of resentment. Why was he attributing to

laziness what was due to circumstance? He was right, equipment did need to be checked every day, but he was overlooking the fact that we had started the day with back-to-back calls, and I simply hadn't had the chance to finish my checklist. I wanted to say so, but something held me back. Long habit, perhaps.

"Fine," I said, "I'll go start the saws."

I pulled the chain saw out of its compartment, yanked on the ripcord, and watched the chain spin as the motor fired. I revved the throttle hard, wondering what to do about Bryce. His manner was becoming insufferable. I was conscious of a mounting anger. First Graham, then Kyle, then Rory, now this. How many bullies did I have to put up with? The worst of it was, I had never truly stood up to any of them, at least not in an assertive manner. I felt the familiar pain rising in my chest.

Things can't go on like this, I thought. *Something has to give.*

It was Kate who had the advice I needed. The next day after work, I laid out the situation while she listened attentively.

"Can you just be humble about it and admit that you made a mistake?" she asked. "I know you weren't fully to blame, and the way he reprimanded you was mean, but don't put yourself in the position of being someone who can't take correction. You can admit that you didn't check the hose properly, and that it affected them. After that, there's nothing wrong with pulling him aside and letting him know that the way he talks to you really bothers you. Just don't make a huge deal out of it. And don't deny your mistakes."

She was right, of course. I wasn't completely blameless. Though I hadn't had the opportunity to complete my check,

I had looked at the hoses and could have spotted the missing length. Still, Bryce's manner had been excessively heavy-handed, and I knew it was time to address it. That night, I lay awake planning my speech. My forehead felt hot, and my chest was still aching. I woke after a few hours of fitful sleep, climbed into my uniform, and left the house to face the unpleasant task ahead. As I drove, I remembered a verse from Matthew's Gospel that had been read at Mass a few days before:

"If your brother sins against you, go and tell him his fault, between you and him alone. If he listens to you, you have gained your brother. But if he does not listen, take one or two others along with you, that every word may be confirmed by the evidence of two or three witnesses."

There was great wisdom in these words, and it was interesting, I thought, that the fire department model for conflict resolution had been built along similar lines. According to our protocols, I must first meet with my opponent alone. If the conflict doesn't resolve, I must bring in another co-worker, preferably an officer. If things still don't resolve, the situation would be reported to the chief. As straightforward as this approach was, there was an empty feeling in my stomach, as if I were going to step off a cliff. I was about to break with a long-standing pattern, and it was not comfortable. My hands shook a little as I pulled into the station parking lot and shut off the motor.

Bryce was putting his gear on the truck as I entered the bay. I walked up to him with determination.

"I need to talk to you," I said, slightly breathless.

He looked up, surprised.

"Okay," he said.

"First of all," I began, "I want you to know that I respect

your years of experience, and I appreciate you getting me up to speed. But I'm having issue with the delivery."

He took a slight step backward. I realized that I was standing too close.

"I can't learn this way," I went on, gaining momentum. "Maybe this is how you treat each other at Station 2, but it's not going to be acceptable with me anymore."

"Wait a minute!" he broke in. "Leave Station 2 out of this! The issue is my delivery. Go on."

"Fine," I conceded. "We'll leave your station out of it, but I'm not very happy about the way I'm being treated. You've got me all wound up, man!"

I had been increasing in volume and speed. It wasn't just him I saw in front of me. All the pent-up frustration with Rory, Graham, and Kyle was spilling out. Bryce regarded me for a moment coolly. He was surprised, I could tell, but not cowed.

"Don't get wound up," he said evenly. "Look, try not to take things so personally. Captain O'Connor asked me to run the station until we have a permanent captain, and so I've been speaking up when I see things that aren't right."

"Sure," I said. "And you were within your rights to correct me last shift. I just don't need a heavy-handed lecture."

"All right, I'll keep it short next time, how's that?" he said with an expression of annoyance.

"That would be good," I replied. There was a pause. It was a concession, but I still wasn't satisfied. There was a tense silence.

"How can we go about this?" he asked finally. "What can I change to make things work?"

"For one thing," I said, "take a look at how Eddie treats me. When things go wrong, we talk about it and move on."

At that moment a door banged, and someone entered the

bay. There were footsteps, the door banged again, and there was silence. I lowered my voice and continued.

"Maybe you don't realize how you're coming across, but you're on your high horse, you're talking down to me all the time. I've had enough!"

I stopped, the emotion spent, my supply of words used up. He returned my gaze steadily, but I noticed the color had drained from his face.

"Okay," he said simply. I was beginning to be impressed by his self-possession, and a little ashamed at my own heat.

"Look," I said. "Don't get me wrong. I'd like to learn from you and work with you. I just hope we can be a little more respectful around here, that's all."

"All right," he nodded. "So let's do that."

Now we had to spend the next twenty-four hours together, and little was said in the awkward aftermath of the confrontation. The release of strong emotion after a sleepless night left a strange lethargy in its wake, but my chest pains were gone. Just after 3 in the afternoon, we were dispatched to a report of a pedestrian run over by a vehicle. As I drove around the S-curve to the east of the station, dispatch came over the radio:

"Engine 6, be advised, this is for a child struck by a vehicle."

My heart beat faster, and I stomped harder on the gas pedal, blaring the air horn longer than usual at each intersection. The truck ground to a halt as we arrived on scene. Police cars were blocking the lane, and a crowd had gathered by a silver SUV. Bryce and Eddie grabbed the medical gear, and I followed at a run. A slender boy about five or six years old was sitting on the ground, holding his leg and fighting back tears. His family members were clustered around him, speaking rapidly in a language I didn't recognize. Bryce squatted down and began to talk to him soothingly, while Eddie pulled back the pant leg to check for damage. The leg did not seem to be broken, but there was a nasty cut on the shin.

"What happened?" asked the captain.

"The SUV was just backing out," the police officer replied. "He didn't see him, backed right over his leg."

"How is it not broken?"

"No idea. He's extremely lucky."

The three of us worked to bandage the leg, and in minutes, the ambulance had arrived and paramedics were taking over. As I stood up and took a step backward, I felt tears coming to my eyes.

Don't be stupid, I told myself severely. I had never experienced emotion on a medical call, and besides that, the kid was all right. Something had changed in me, and I didn't like it. Maybe it was the fact that the boy was around the same age as my own daughter; maybe it was because of my conversation with Bryce earlier. Rory's words came to mind. *When you feel the tears coming to your eyes on medical calls,* he had said, *it's time to quit.*

No way, I thought to myself. *It's too soon. I've only been on the job five years!*

That was the first call of the day. Not long after, we were dispatched to a serious MVC at a busy intersection. A four-door sedan had run a red light and caught another vehicle in a vicious T-bone. One car was crushed up against a light pole, and the other sat sideways to oncoming traffic. The occupants of both vehicles were out and standing on the edge of the roadway. They looked badly shaken. I parked with the truck angled to protect the scene, cutting off at least one lane of traffic, and began laying out cones to prevent cars from accessing the intersection. Bryce and Eddie headed quickly to the patients, while I directed traffic around the accident. As I was waving cars past, a taxi suddenly veered out of its lane, cut across the intersection and stopped in front of me. The rear window rolled down and a young lady looked out at me.

"Can you let me through?" she asked sweetly. "I need to get to the train station, and my train leaves in five minutes."

"I'm sorry," I replied. "You can't turn in here. This is an accident scene. You'll have to cross the bridge, take Alexander and come back on Bantry. There are signs for the train station. You still might make it."

"Please," her tone got sharper. "I'm going to miss my train!"

"No!" I answered, my voice rising. "I can't let you through!" I was surprised by my own vehemence. I had never yelled at a member of the public before. The taxi driver cut in.

"Just move your cones," he said bossily. "There's room for me to get by your truck. The station's right over there."

"No!" I snapped back at him. "You have to move on! You're going to cause another accident sitting here!"

With an exasperated expression, the woman rolled up her window, and the car swerved back into traffic, causing another vehicle to brake and sound its horn. I felt suddenly guilty. First tears, now losing my temper! What was wrong with me?

The next shift, I drove in to work with the familiar fire in my chest. On afternoon break, I picked up the phone and dialed the number for Station 4. The cheery voice of Chuck Brown answered.

"Hey, Captain, it's Ben."

"Big Ben! How are things, man?"

"Okay, sir. I just wanted to run something by you."

"Sure, go ahead."

"Look, uh, I think I'm having some stress issues here."

"What kind of stress issues?" he asked.

"Chest pains," I said.

"Are you serious? You gotta get that checked out!"

"Oh, I have. They said its nothing to worry about, just a little side effect of stress. I've talked to someone from the mental health team, too. It's all been very helpful, but I'm not sure what to do next."

"Do you want out of there?"

"No-o," I said slowly. "I really have no reason to. The two people I was having problems with aren't causing me trouble any more."

"Well, look, if it gets too much, let me know, okay? I'll talk to the chief, and he'll have you out of there in a heart-beat."

"I will, Captain, thanks."

"And don't hesitate to call me again if you need someone to talk to."

"I appreciate that, thanks."

I hung up, feeling more uncertain than ever. Would a change of stations really make a difference? Maybe the problem was me. Maybe I wasn't cut out for this job. Why did everyone around me seem so confident? Maybe it didn't matter where I was stationed. I would only bring my problems with me wherever I went. Maybe it was time for Chuck to know the full story. I had just got off the phone with him, but I would be better able to explain things if I wrote them out. I sat down at the computer and began composing an email.

Chuck,

I want to be honest about my real reasons for leaving Station 17. I don't think you saw much of it during your short time there, but I left because of the way I was being treated. I thought that if I could go to a busy station there wouldn't be time for harassment. I was wrong. At Station 6 I began having chest pains and insomnia. Now, all I want is to work with people who follow our respectful workplace policy, and I would be willing to go to a different shift, Fire Prevention, Training Division, or even another job to make it work.

Sincerely,
Ben

I paused for a moment before sending it. I had a feeling that it would set off a chain reaction, and what the results would be I could not foresee. I took a breath and hit "send."

An hour later, a white chief's car drew up in front of the station, and Division Chief MacLean stepped out. Captain McCannell must have seen it from his office window, too, for a moment later he was walking out onto the tarmac. The two of them stood in conversation for some time. I watched as they came toward the doors. Mike was nodding, and the chief looked serious. We met in the hallway.

"A word with you, please," the chief said simply.

He led the way into the captain's office and shut the door. MacLean leaned against the desk facing me, Mike took the only chair, and I sat on the end of the bed since there was nowhere else. Chief MacLean was a stolid character, not known for his people skills. I couldn't read his expression as he looked me over.

"I saw your email," he began. "Captain Brown forwarded it to Chief Downey, who forwarded it to me. I was astounded, quite frankly, knowing who's at Station 17."

If you knew them, I thought, *you wouldn't be.*

"Well, sir," I began. "What it amounted to was an individual who did not like or respect the fact that I'm Catholic. He succeeded in making my life very difficult for a few years. Then there was someone else I didn't get along with. By the end of my time there, I would say things had definitely improved, and I think if I went back there now and told them to stop, they would. The question is, why am I still having stress issues?"

He stared at me for a moment.

"Next time you have problems, I want you to contact me—immediately," he said. "I put you at Station 6 to get you more exposure to fires, which obviously you weren't getting at 17. I did that knowing I was putting you with a very demanding officer. I didn't know you were having issues."

"I appreciate you giving me a busy station," I replied. "And in fairness to Captain O'Connor, he really toned it down after I told him about the chest pain. In fact, it was he who got me set up with mental health services. At this point, Chief, I would consider any personnel issues resolved. I just can't seem to get a handle on these symptoms."

"You were at the O'Donnell Street fire, weren't you?" he asked quietly.

"Yes," I replied.

"Is it the calls then?"

"No," I shook my head emphatically. "I want to be very clear about that. I still love what we do. I love going to fires, I even enjoy driving now. To be honest, it's the culture we have in our stations, and the way we treat each other sometimes."

He nodded.

"We've all been through it," he said. "I've had to put in some years under officers who were real dinks, for lack of a better term. Keep in mind that we have a chain of command. People move up, your turn will come, but you may have to put up with people who are really difficult to get along with—for a time."

Mike was nodding hard at this, with an expression that said, *yep, I've been there too*.

"I don't know about Fire Prevention," the chief went on, "but I can get you a transfer to another station if that would help. Is there somewhere you want to go?"

"Can I take some time to think about it?" I asked.

"I'll give you a week. Send me an email next Friday with two or three stations you're interested in, and I'll see what I can do."

He moved toward the door.

"Thank you, sir," I said, standing.

"If you ever need to talk, you can contact me," he said gruffly.

Yeah, right, I thought.

A few days later, I was detailed to a west-end station for one shift. The captain and crew were new to me, and although we made small talk, I kept my situation to myself. The morning was taken up with calls: a small bedroom fire, medicals, and an MVC. Midway through the afternoon, the captain came into the kitchen with a puzzled frown.

"I just got a call from the District Chief," he said. "Apparently, there's some top brass coming down from headquarters. Says he wants to talk to you, Ben."

The other two firefighters looked up in surprise.

"What did you do?" one of them asked.

"Oh, I think I know what this is about," I replied. "I was

having trouble with some of the guys at my old station. I went to someone for advice. Chief got wind of it, and now it sounds like things got blown out of proportion."

The other firefighter at the table belonged to A Shift and was working overtime that day. He shook his head.

"You get all the bullies on C Shift," he said. "It's no secret."

I left them to find their button-up shirts and shine their boots, while I headed to the bay floor to pull myself together.

You've got nothing to be ashamed of, I told myself. *Just be honest, and don't play the victim.*

There were butterflies in my stomach as I watched the staff car pull up. A tall, slender man in his late forties got out. He wore a white shirt and had four gold bars on each shoulder: a Deputy Chief, second in command of the entire department. He carried himself with a natural authority, but there was an openness to his face as we shook hands.

"John Forbes," he introduced himself. "You must be Ben."

"Yes, sir," I said. "Thanks for coming down."

"Oh, you're welcome. You'll have to excuse me landing on you like this. It won't take long."

"That's quite all right, Chief."

The crew was assembled in the hallway, and after a short greeting the chief said:

"I just need to steal this young man for a few minutes. Can we use the training room?"

The captain gave his assent, and once the door was closed we sat facing each other across a small table.

He produced a portfolio with some papers. As he was arranging them on the desk, I saw him touch something black behind the flap. Perhaps he was only silencing his phone, but I wondered if I was being recorded.

"I hope you don't mind my coming to see you," he began. "My intention is certainly not to intimidate you. Your Division Chief brought up your case to me at a high-level meeting, and I thought I should meet with you myself. As you can appreciate, we take it very, very seriously when we hear about workplace harassment and violence."

"Thanks, Chief," I replied, a little alarmed that he had used the term "violence." Had my case been exaggerated?

"Rest assured," I said hastily. "There's been no violence. Some verbal harassment maybe, that's all."

"Well, it's no secret that your captain was a very difficult person to work for," he remarked.

"Yes, it was difficult," I admitted. "It's not that he did anything wrong. He just had a very aggressive management style, and I guess I'm not as resilient as the next guy. Then, as far as what happened at 17—"

"I don't know any details," he broke in. "I appreciate your honesty, but don't feel you have to disclose everything."

"Okay," I said. "Without naming names, then, there were a couple of gruff characters at 17 who went out of their way to give me a hard time. As I told Chief MacLean, I know they would stop if I told them to today. I know those guys. But I was new, and I thought I had to prove that nothing bothered me. It didn't work. I would consider those conflicts resolved," I went on. "But what's not resolved is my mental state."

"Have you accessed Support Services?" he asked.

"Yes, I have. The person I've been talking with has been very helpful. I just don't think anything is going to change while I'm at Station 6."

"Is it the calls?" The same question MacLean had asked.

"No," I replied. "I still love this job. No, it's what goes on in our stations."

"Rest assured," he said, "that from a management perspective, we are working very hard to eliminate the problem. It may take years, though. So, thank you for being part of that change. . . . Now, your Division Chief offered to move you. Do you think that's a good solution for the time being?"

"I think so," I said. "I've come up with some stations where I know the guys, and the officers seem pretty decent. I was going to submit my request on Friday."

"Good. Well, I've heard from more than one District Chief that you're a hard worker. We want to look after good employees."

His finger strayed to the black device. He closed the lid and gathered up his papers. We stood.

"You take care of yourself," he said. "And listen, if you want someone to talk to, call me directly. A Deputy's always busy, but I'll make time for my people. Don't hesitate."

We shook hands.

"Thank you, Chief," I said. "I appreciate it."

I knew he really meant it.

13

IMITATION

The first rays of the morning sun were peeking over the roofs of Lower East Side as I pulled up to Station 8 for another day at work. I couldn't help feeling a little surge of anticipation as I thought of the day ahead. Tyler, our rookie, was already in there working hard; I could see his car in the parking lot. Clay, the senior man, would be showing up in about twenty minutes, fashionably late as usual. Our captain was already in the kitchen, I could tell. I could hear his booming laugh all the way out here. It must be an uproarious conversation he was having with the off-going crew.

Thank you, Lord, for Station 8, I prayed quietly. It was something I had found myself saying often in the past year and a half.

I walked onto the bay floor and saw Tyler coming around the corner of Engine 8.

"Morning, Ty," I greeted him cheerfully.

"Hey, man!" he replied enthusiastically. "How's it going?"

"Good," I replied. "Looking forward to another day at the Great Eight."

He grinned.

"Med gear's all checked," he informed me. "Extrication equipment is good, nozzles are good. I'm going to get some eggs on in a minute if you'd like some."

"Count me in," I said. "I'm hungry. And I'm sure the captain will be, too."

I watched him disappear down the hallway in the direction of the kitchen. Tyler had been a forest-fire fighter before getting hired with the city and had brought his strong work ethic with him. Landing at Station 8 after his probationary year, he had assumed his role as junior man with enthusiasm. Already, the smell of coffee was wafting through the station, and a quick glance outside told me he had hoisted the flag. There wasn't much for me to do on the truck, since he had gone through the main equipment already. I arranged my bunker gear and helmet, checked my portable radio and SCBA, then headed to the locker room to stow my kit bag. Photos of Kate and the girls smiled at me from my locker door.

I stowed my gear and headed to the kitchen. The B Shift personnel had left, and the captain and Tyler were animatedly discussing sports. I could hear the captain's voice echoing loudly down the hallway. Frank, who had been my acting captain at Station 15 during my probationary year, had been assigned a permanent spot at No. 8. Cheerful and large as ever, he looked up as I entered.

"Hey, mornin' Ben!" he hailed.

"Morning, Captain," I replied. "How was your time off?"

"Good, brother! How about you?"

"It was good. Got a bunch of work done around the farm."

"Right on. Can't say I got much work done," he laughed. "I took the motorbike out for a spin, though. Man, yesterday was a beautiful day!"

"What kind of bike you got?" asked Tyler, turning from the stove with spatula in hand.

"It's a Harley," he replied. "Road King. Got the 1690 cc engine in it . . . it's a good ride."

"Nice!" said Tyler. "That's my next bike. I bought a 2011 Street Glide. It'll do for now."

At that moment, Clay walked in. He was portly, with a blond comb-over.

"If you two are going to start talking motorcycles," he said dryly, "I'm going home."

Frank's laugh rang out again.

"Morning, Clay," he greeted. "You're just in time for some eggs."

"Yuck," Clay replied. There was a twinkle in his eye as he slapped Tyler on the shoulder. "Just kidding. How's our fearless rookie this morning?"

"I'll be even better in two minutes," said Tyler. "These are nearly done."

He began spooning generous portions of scrambled eggs onto four plates. I stepped over to the sink and began washing his prep dishes.

"Hey, leave those alone!" he scolded. "That's my job."

"I don't mind," I said, smiling. I made it my policy always to be kind to rookies. "You've been going non-stop this morning."

A few minutes later, we were sitting down to breakfast. Frank swallowed his eggs with gusto, exclaiming, "Dang, these are good!" between each mouthful.

"They're all right," said Clay blandly, the twinkle showing in his eye again.

"How's your arm doing, Captain?" I asked.

"I'm going for another X-ray next week," he answered. "Something's not right in there. It still hurts when I twist it."

"You're lucky it was just an arm," said Clay. "Crazy driver like you."

"Heck, I've broken lots of bones," exclaimed Frank. "There was my collar bone three years ago, that was taking a spill off the snowmobile. Then there was the time I put the bike down on a patch of gravel, that cost me two ribs. I've been knocked out more than once, too!"

"That's why I don't like getting on anything with less than four wheels," I remarked.

"Come on man, you gotta live a little!" he cajoled. "And anyway, I was on four wheels! It was my ATV. As long as you get a good helmet, that's what's important. Mine cost me fifteen hundred bucks. Like I always say, you got a thirty dollar head, buy a thirty dollar helmet. And that reminds me . . . " But his words were cut short by the station alarm. Tones for a medical. Forks were dropped, chairs scraped, and we hurried toward the bay floor. I turned into the watch room, grabbed the printout, and keyed the mic.

"Engine 8, we're going to 2237 Sunfish Crescent, for a 23-year-old male unconscious. Possible overdose."

I ran to the truck, hit the button to open the garage doors, fired the ignition, and climbed into the driver's seat.

"Where's Sunfish?" I called out to Clay, who was just zipping up his bunker jacket.

"It's off of Magnolia, right before the high school," he shot back. "Just go, I'll get it on the GPS."

Doors banged. One last turn of the head to check that everyone was seated and ready, then I flicked on the red emergency flashers, released the park brake and hit the gas pedal. Engine 8 rolled forward. I swerved northbound onto 18th Avenue, sounding the air horn as cars pulled to the right to let us pass. At the roundabout, I merged onto Magnolia heading West and saw the high school looming four blocks down.

"Keep going," Clay instructed. "I'll let you know when to turn."

The light at the intersection ahead went red. I slowed to a crawl, sounding long intermittent blasts on the horn, and the vehicles waiting at the stop line in front of us began easing right and left to allow us room.

"Red light camera!" Frank warned. I came to a complete stop. *Left, right, left again,* I thought as my head turned back and forth, scanning the intersection for traffic. A black minivan and a silver car had stopped to let us through. My foot came off the brake, and we began to crawl forward.

"Whoa!" yelled Frank. I braked again. Another vehicle had moved up in the lane beside the minivan and was racing through at top speed. I blared the horn, but he swept by without heeding.

"Idiot!" Frank snorted. "Not you," he added quickly. "Some drivers just don't pay attention."

I stomped on the gas, and we accelerated through the intersection. The next two lights were green, and I drove on at top speed.

"Right turn coming up!" called Clay. I slowed at the corner and spun the wheel, bringing us onto Sunfish Crescent.

"After the cross street, it's the third driveway on the left!" came the next instruction. I slowed. We passed the four-way stop. There was 2237, a large, two-story home, of recent construction. The door was open. I parked two houses down to leave room for the ambulance. Clay and Tyler hurried out of the cab with their medical equipment, as the captain gave a radio update. I could hear another siren in the distance. The paramedics weren't far behind us. As the driver, my job would be to stay on the street and assist them, unless the captain needed an extra hand inside. I watched as the ambulance rounded the corner and came to a stop in front of the address. Two male paramedics, one young and one middle-aged, got out unhurriedly and began unloading their stretcher.

"What have we got?" the older one asked me.

"Crew's just inside assessing," I answered. "Unconscious male, possible overdose, is all we know so far."

"Us too," he said. I helped him wheel the stretcher up the pathway to the door. He gave me a nod before he and his partner disappeared inside. I waited, two minutes, possibly three. I could hear what sounded like muffled yells coming from somewhere inside. It wasn't long before Frank's voice came over my portable radio.

"Engine 8 Operator from Sunfish Command."

"Go ahead," I replied.

"Can you don some medical gear and give us a hand?"

"Copy," I responded, hurrying to the truck for my gloves and safety glasses. Entering the house, I saw a young man, pale and emaciated, struggling to tear the oxygen mask off his face. Clay, Tyler, and the younger paramedic knelt beside the patient trying to restrain him. His motions were getting more frantic by the second. A middle-aged man who I took to be the father hovered nearby, his face registering a kind of pain that only a parent can feel. Two nasal injectors, with "Narcan" written across each of them, lay on the floor. It was the new antidote used to treat opioid overdose, and we carried it in our med bag. It could wake a patient up from drug-induced unconsciousness very quickly, but the results were often unpredictable. This patient did not appear to be coping well.

"We're going to move him to the stretcher. You and Ty grab his upper body," Frank ordered. "And pin his arms. Clay, you help our medic friend lift by the legs."

Tyler and I took our positions by the young man's head. We each slipped an arm around his back, crossing our arms behind him, and with our free hands grabbed his struggling wrists. He was thrashing and twisting now, trying to break free. I was surprised at the sheer strength in such a frail body.

"Take it easy buddy," I coached, gripping tighter.

"On three," said the medic. "One, two, *three*!" We lifted. The patient started to scream, arching his back like it was touching hot coals. We muscled him onto the stretcher, and I fought to keep him down as the other three clipped the straps over his legs, chest, and pelvis. I managed to pin his shoulders long enough for them to tighten the straps down. After a final struggle and a desperate scream, he seemed to realize that he couldn't escape, and he settled down to a quiet moaning and shivering. We quickly wheeled the stretcher to the waiting ambulance, lifted it together, and slid it inside. The senior medic climbed in beside the patient.

"Which hospital?" his colleague asked.

"Central," the other replied. The doors banged shut. A moment later they were driving away, and as their siren faded in the distance, we disposed of our gloves in the onboard waste container, returned the medical supplies to their shelf, and mounted our own truck.

"That was a nice shack," observed Tyler as we drove. "Nice neighborhood, too. Not the kind of place you'd expect to get an overdose call."

"Ah, that's the fentanyl crisis for you," said Frank. "It's everywhere now. Rich, poor, doesn't matter."

"You gave two doses of Narcan?" I asked.

"No, that wasn't us," said Tyler. "The Dad said he gave him two shots. Must be an ongoing problem if he's got Narcan in the house."

"I feel bad for him," I said. "I can't imagine what it must be like seeing your kid go through something like that."

"You gotta wonder how long it's been going on, too," put in Frank. "In rehab, out of rehab. Jail maybe. At some point you just throw up your hands and say you've done everything you can for him."

Suddenly, Clay's voice burst out in full-throated song. It was a recent hit that was playing on the radio, and he was slightly off-key. Frank's laugh boomed through the truck. I grinned in spite of myself. Tyler shook his head.

"You are something else, Clay," said Frank. "Way to change the subject."

We took a quick detour to the fuel yard to put diesel in the truck. Next stop was the supermarket, and Tyler and Clay went in to buy our groceries for the day. Back at the station, Tyler and I re-stocked the medical equipment, while Clay whipped up an impressive batch of salmon cheese melts with tortellini soup. We sat down to an 11:30 lunch.

"Dang, that's good!" exclaimed Frank between mouthfuls.

"It's all right," said Clay blandly, the twinkle showing in his eye.

After lunch, Tyler washed dishes as Clay and I dried and tidied up. Kitchen chores done, Tyler headed to the watch for his stint on office duty, while Frank and Clay disappeared for their two hours of free time.

"*You and I,*" sang Clay, on-key this time, as he retreated down the hall. Frank's laugh followed him.

"No, just you, buddy," he retorted. "I'm going to study!"

"Study" was code for taking a nap. Our rules didn't allow us to sleep on duty until after 10 o'clock at night, but most officers allowed a two-hour "study time" after lunch.

"You sure you don't want me to take watch for you?" I

called after Tyler. "You can go study if you like. I can hang out there and babysit the phone."

"No way, man," he called back. "Rookie job. You go relax."

I waved my thanks and headed toward the dorm. I had done watch duty every afternoon for my first five years, and now I was afforded the luxury of a little down time after lunch. This had become my prayer hour. I usually found a quiet corner of the station where I could read my Bible undisturbed and spend a few minutes in silent meditation. Kate had given me a new book to read called *The Imitation of Christ.* I had taken only a cursory interest in it at first. To be honest, it had looked like just another dry theological tract, but I had brought it along partly to please her, partly out of curiosity. However shallow these motives, I began reading, and it wasn't long before I became completely hooked. Whoever this Thomas à Kempis was, his insights into human nature were incisive. Even though he was writing over 500 years ago, it was as if his words were aimed directly at me and my problems in the modern fire station. A passage leaped off the page at me:

> My son, stand firm and put your trust in Me; for what are words, but words? They fly through the air like mere pebbles that cannot hurt. If you are guilty of the accusations, consider how you would gladly amend yourself. If your conscience does not reproach you, consider that you would gladly suffer this for God's sake. It is little enough to sometimes suffer from words, since you do not yet have the courage to endure hard lashes. And why do such small matters go to your heart, unless it is because you are still carnal, and you regard men more than you should? For it is because you are afraid of being despised that you are unwilling to be reproved for your faults, and so you seek the shelter of excuses.

I put the book down. He had spoken directly to one of my greatest struggles. Memories from Station 17 and Station 6 floated across my mind. With them surfaced the old frustrations. Sometimes when I thought about Kyle, or Graham, or other bullies, a sickly sweet fantasy played in my head. I imagined myself blasting them with righteous anger, cutting them down with withering rhetoric. In my most vivid scenes, I would even picture myself in a burst of shouting and foul language. Yet, as I fed these imaginings, the burden of anger never lifted. Only if I caught myself and made a conscious act of forgiveness would the feelings recede. Shaking these thoughts away, I read on:

> Do not say, "I cannot endure to suffer these things at the hands of a this particular person, nor should I endure things of this sort—for he has done me great wrong, and he reproaches me with things which I never thought of; but I will willingly suffer at the hands of another, that is, if they are things I see I ought to suffer." Such a thought is foolish; it does not consider the virtue of patience, nor by whom its crown is received. Rather, it weighs too closely the persons and the injuries from which it suffers. One who is willing to suffer only so much as he thinks good, and from whom he pleases, is not truly patient. Rather, the truly patient man does not mind by whom he is abused, whether by his superiors, by an equal, or by an inferior; whether it is by a good and holy man, or by one who is perverse and unworthy. But however much, or however often anything adverse befalls him, he takes it all thankfully, and indifferently from every creature, as if it were from the hands of God; he considers it a great gain. For with God, it is impossible that anything suffered for God's sake, however small, will come to pass without its just reward.

I could only laugh ruefully at myself. It was as if he knew me. I thought of my confrontation with Bryce. I had been rather proud of myself in the aftermath. I now saw how attached I had been to my own hurt feelings. Perhaps it was time for the next stage, detachment. Who knows what other strong characters would try to cross boundaries in the years ahead. I had learned how important it was to stand up for myself; in future, I also needed to let go of my own ego.

"Okay, Lord," I said. "I would like to put these words into practice." *Still,* I thought, *please don't take this prayer* too *literally.*

At 2 P.M. we started our daily training block, and for the next two hours, we practiced raising ladders against the side of the station. After training, I headed to the kitchen and began prepping the evening meal. Tyler went into the gym, and soon I could hear his heavy metal music blasting, along with the occasional clunk of weights. Clay was watching baseball. I chopped onions and peppers and set them to fry in the wok. The captain wandered in.

"What's on the menu?" he asked.

"Something new today, Cap," I responded. "Thai chicken."

"Oh, sounds good."

He wandered away again. I began treating the chicken breasts with a coating of olive oil and Montreal chicken seasoning. I covered the bowl and started the rice to boil. Presently Tyler emerged in a sweaty T-shirt and asked:

"Need a hand?"

"Go shower," I said. "I'll get you to put the chicken on the barbecue in about fifteen minutes."

An hour later, we were sitting down. Frank took a mouthful and exclaimed:

"Man! That is good! What did you say it was again?"

"Thai chicken," I replied. "It's a new recipe I learned. First time trying it at work."

"What's in it?" he asked.

"It's actually really simple. Just peanut butter and salsa mixed in with the chicken and veggies."

"Really? That's it?"

"Yeah!"

I waited for Clay's inevitable, "it's all right," but instead he chewed thoughtfully, then pointed at his plate with his fork.

"I don't say it very often," he said with his mouth full. "But this is very good."

The talk strayed onto other topics, and midway through the meal I noticed Frank looking at his cell phone.

"Ooh, she's hot!" he remarked, holding out his phone so that Clay could take a look. Tyler craned his neck also and whistled.

"Ben?" he inquired, reaching over for me to see. *Here we go,* I thought. *Trouble.* But my experiences at other stations weren't for nothing.

"Nope, I'm good," I answered confidently. "I don't look at that stuff."

"Aw, come on!" he threw his hands up. I merely smiled in response. He shrugged and went back to his meal, and the conversation moved on.

What are words, but words? I reminded myself. *They fly through the air like pebbles but cannot hurt anyone.* An image, on the other hand . . .

~

It was evening. Dishes were done, the station was clean, and the flag was down.

"You going to play your fiddle for us tonight?" asked Clay. "We're overdue for a concert."

"That's the plan," I answered. "I brought it in today."

I went to my locker and lifted out the violin case. I took up a chair in the kitchen, lifted the instrument out carefully, and plucked the strings gently to check for tune. I tightened the bow, ran the horse hairs up and down over a lump of rosin, then placed the fiddle under my chin. After some warm-up notes, I began sawing away. I played *Drowsy Maggie,* followed by *Swallowtail Jig*, then the *Rocky Road to Dublin*, and finally *McGovern's Reel.* There were a number of stumbles, but I recovered and kept going. Clay sat nearby, taking it all in with a little smile. I was partway through *Devil's Dream,* when there was a curious clicking sound coming up the hallway. The captain came round the corner, arms folded across his chest like a Russian dancer, his heels shooting out right and left to strike the floor in time to the music. The sight was so comical I lost control of the bow and couldn't finish.

"Keep going, don't stop!" he said, leaning over to pant.

"You threw me off," I laughed.

He turned and retreated into his office. I played "Loch Lomond" to wind down, then put the fiddle away.

It was time to wash the truck. We sprayed, soaked, and scrubbed it well and dried the bay floor with long-handled squeegees, the last chore of the day. It was 10 P.M., time to turn in for the night. As the driver, I slept on the small cot in the watch, ready to jump for the printer as soon as the call came in. That night, however, there was nothing, and I drove home the next morning to a bright sunrise feeling rested. I smiled as I thought of Frank dancing in the kitchen. *What a difference a good crew makes*, I thought. Life had certainly changed a lot since coming to No. 8. Things weren't perfect, but an awareness was dawning that I was beginning to love my job.

The next shift, I arrived a little earlier than usual. Tyler wasn't there yet, and I had a moment with Frank alone.

"Can I talk to you about something?" I asked.

"Sure, buddy, what's up?"

"You know that pinup girl in the gym?"

"Yeah."

"Well, I've been thinking. It's not very appropriate for the workplace, and we have female firefighters here on other shifts. We should probably take it down, shouldn't we?"

I had decided on this as the best angle. I wished I could give him a crash course on the *Theology of the Body*, but I knew I needed to meet him where he was at. I had been covering the picture with a workout mat whenever I was in the gym, but finally I had resolved to do something more decisive about it.

"Yes, I suppose we need to be respectful," Frank said thoughtfully. "Just put it in the electrical closet for now."

I did as he directed, but a few shifts later, it was back on the wall. I waited until "study time", then ripped it up, stuffed the remains in a garbage bag and threw it in the dumpster. The others still had their phones, but now publicly, at least, a woman would not be on display as an object.

The following shift dawned bright and sunny. It was the in-between time after the snow had melted and before the grass had started to turn green. The ditches, parkland, and fields in our district lay brown and dry in the spring sunshine.

At a quarter to eight we heard the tones, and Clay's voice over the speaker:

"Grass fire! Near the water tower!"

Moments later we were on our way in a howl of sirens and a blaring of horns.

"How do we get to the water tower?" I asked from behind the wheel.

"Follow 17th Avenue, then turn right onto Owen's Park Road," answered Frank. "It's that big patch of trees right outside the subdivision."

"That's a big area," said Clay. "There's a good thirty acres in there."

We turned onto Owen's Park. Beyond the last row of rooftops lay a wide belt of forest, with a curtain of smoke rising above it. Two fire trucks were staged at the side of the road, and hoses stretched into the trees. I parked behind them, and we exited the truck.

"Forestry equipment," ordered Frank. We opened various compartments, pulling out long-handled flails, shovels, and a portable water tank with hose and nozzle that could be strapped to one's back. The forestry bag followed, our most important piece of equipment. It weighed nearly a hundred pounds and contained four lengths of hose, a nozzle, and various tools. Tyler shouldered this heaviest load, and we trudged up a gravel walking path that led into the parkland. As we came out of the woods into a wide open space, a line of fire was creeping in our direction, devouring the dry grass as it came. Firefighters were beating at it with flails and spraying it with their pump tanks. An officer approached us.

"We need more reach on our hose! Can you guys hook into our line and try get around the flank? It'll be in the trees in a minute!"

We scurried to obey. Clay grabbed the end of the hose out of Tyler's backpack and attached it to the hose from the first-arriving engine. Tyler took off at a jog toward the flank, the hose sliding out of his bag as he ran. I followed with the shovel and flails. In a moment we were in the trees. The fire had not reached us yet. The last of the hose flew out of the pack, and I grabbed the end that had the nozzle attached. A second later, the line stiffened as it filled with water. The flames were just licking the bottom of the trees. I aimed a blast, and a patch of fire hissed and died. I walked along the edge of the fire line, swirling the water stream in a circular motion, simultaneously extinguishing the surface fires and digging into the soil to reach the smoldering roots. Tyler followed with the shovel, stamping out any remaining embers. We rounded the flank and stopped short when the hose reached its full length. The fire appeared to be out, at least on this side.

"Engine 8 crew from command!" crackled on my portable radio.

"Go ahead."

"Bring your line across to the north side of the fire!"

We gathered up the charged hose line as best we could and began trudging in that direction. There was a lot of hose, with many loops and turns among the trees, and it kept snagging and slowing our progress. We had just emerged onto the blackened open space, when a red-helmeted figure strode toward us. For a second, I half-thought it was Rory, for the eyes had the same commanding intensity, but then I realized it was someone I had never met before.

"*Get that hose line across*!" he yelled at us. There was an edge of panic in his voice. He pointed over the open space to another line of trees at the north end of the clearing. "*Go*!"

We rushed across, clouds of ash swirling around our ankles. The occasional hot spot spurted smoke. The fire in the open space was out, but a section of it must have got away and run into the trees. As we neared the forest, I could hear the crackle of burning underbrush. I crashed through the overhanging branches and came to the fire line. It wasn't huge, and it was moving slowly, but it could get out of control if we didn't get water on it in a hurry. I aimed the nozzle and opened it, but only a sluggish stream came out. In a moment it had slowed to a trickle, then stopped.

There was a crunching in the bushes behind me, and Tyler emerged.

"Your hose burned through," he said. "Give me a minute, we'll get a new length. Another crew's helping me."

I waited, watching the fire creep farther and farther into the woods. A minute went by, then another, and another. What was taking them so long? I turned around and walked out of the trees, still holding the nozzle. Perhaps they had forgotten about me and were deploying another line. The officer who had yelled at us was standing beside a chief, watching the fire. A crew was just tightening the last couplings on a new length of hose, attaching it to mine.

"You still want my line?" I called. "I'm ready for water any time."

"Yeah!" the officer rapped, "and you can get back to the fire! *That* would be a good idea." He rolled his eyes and gave a smirk of contempt. For a second I thought of retorting, but then I remembered that a fire scene is no place for an altercation. I returned to my position on the fire line, opened the nozzle, and began attacking the fire. In minutes, it was out. We worked for another hour, walking around the edge of the burnt-out area, finding hot-spots and digging up roots that were burning beneath the surface. The fire had devoured about ten acres of grassland, but we had saved the forest. Chain-saws buzzed as crews cut down a few charred trees where it had encroached into the woods.

We gathered up burned and blackened hoses, found tools discarded in the heat of the moment, and gradually made our way back to the truck. Everyone's faces were grimy, and lines of weariness showed around the eyes.

"Engine 8, from Command, you can clear," came the welcome order on Frank's radio.

"Back to the city," he said cheerfully. Soon we were driving out of the forested area in the direction of the suburbs. Clay belted out a country song off-key, something about dirt roads.

Back at the station, we had a big cleanup job to do. The burnt length of hose needed to be replaced and the other sooty lengths washed and hung in the hose tower. Our dirty bunker gear had to be bagged and placed by the front door. A maintenance crew would pick up the bags and bring them to headquarters for cleaning sometime later in the week. The city had recently issued every firefighter a second suit, which was kept hanging in the gear room, so we didn't need to drive down to headquarters for a fresh set. Still, all the tools had to be cleaned and inspected, the backpack pump tank refilled, and the forestry bag repacked with fresh hose. It took about an hour, but finally the truck was response-ready again.

"Can we have a discussion about the fire?" Tyler asked, when we had showered and changed into fresh uniforms.

"Sure," said Frank. "Let's meet in the kitchen."

We gathered around the table.

"So, I know I'm the new guy," Tyler began. "But coming from the forest-fire world, I have some frustrations about how it went today."

"Okay," said Frank. "Go ahead, let's hear them."

"First, I think we could really use some practice deploying the forestry backpack. We had our hose all tangled up in the trees, and when we needed to move in a hurry, we ran into problems."

"Sure, we can do that," Frank agreed. "A training session on fighting bush fires would be good for all of us. Heck, it's not like we get them that often here."

"The other thing is, we would never take a hose line across the burnt area like that," Tyler went on. "As you saw, it ruined our line and caused a real headache. And the officer who told us to do it was very aggressive."

"Yeah, who was that?" I asked. "Maybe you know, Captain. Some guy in charge of Engine 15. He was quite rude."

"Engine 15?" Frank replied. "Oh that would have been Alex Gerard. They call him 'Commander Alexander'. He's loud, but he's harmless. If he does it again, come to me. I have no problem telling off other officers. I've done it my whole career. I've even done it to chiefs!"

He turned to Clay, who had been quietly taking in the conversation.

"What about you, Clay? Anything to add?"

"No," he said complacently. "I thought it went well. We got the fire out, and nobody got hurt."

Frank laughed.

"There's the incorrigible optimist! All right, guys. Thanks for the feedback. You all did a good job, and we'll train some more and work out some of these details."

Alex Gerard, I thought. *I'm going to phone up Station 15 and give him a piece of my mind.* Then I caught myself. I had wanted to put the principles from the *Imitation of Christ* into practice, and here was my opportunity.

All right, Commander Alexander, I thought. *I'll forgive you.* I was gritting my teeth a little at first, but the more I thought about it, the more I realized that my refusal to talk back to him, far from being a show of weakness, had been an act of professional self-control. With that, I felt my mood begin to lighten.

One opportunity will often lead to another. It wasn't long after when Bryce Halton came up in conversation.

"I was talking to an old station-mate of yours," Frank mentioned one morning. "Mr. Halton. I always thought he was a nice guy, but he started going off about you, and I thought, wow, he can be a jerk."

"What do you mean?" I said sharply.

"Oh, something about you not being the greatest team player. I didn't agree with him, of course, but everyone's got his opinion."

I was seething.

"Of all the . . . " I began, then stopped. It would not do to lose my temper. "I don't think he likes me much," I ventured more calmly. "We had a little run-in at Station 6."

"I thought there might be some kind of history there," Frank observed. "That's too bad. You're both good firefighters."

I made no further comment. So, he had been talking behind my back. Again I entertained the idea of an angry phone call.

Why do such small matters go to your heart? Consider that you would gladly suffer this for God's sake.

Swallowing my pride, I made a small act of forgiveness and went about my daily duties. The next morning, however, the temptation had returned in full force. How was forgiving different from letting myself get walked over? In the past, I had been unable to bring myself to confront such people, even when I would have been justified in doing so. Eventually I had gained the confidence to assert myself. Now, it seemed, I was expected to go backward and become passive again. Did being a Christian mean being a pushover? The turnoff to Station 6 was on my way home, and it suddenly came to me that I could stop in and have a word with him. As I drove, I thought it over. I had been relieved of duty early, and there was a good chance Bryce was still at the station. Choice phrases floated through my mind.

Then something else occurred to me. Something a wise priest had once said. There is a great difference between suffering unwillingly, and consenting in freedom to what one normally would not choose. A lot had changed in the last year. No one could call me a pushover any more. But I had a choice in front of me. I was more than prepared to tell Bryce off, but if I wanted, I could say nothing and offer it up as a kind of sacrifice. Perhaps I could win some graces for the world. It was a freeing thought.

The exit was coming up on the right. My hand strayed to the turn signal.

No, I told myself. *This is not the way to deal with it.* I held the wheel steady and kept going. Maybe another time there would be a chance to challenge him about talking behind my back, but it would have to be done with gentleness. Today the call was to let it go.

Clay booked off sick the next shift, and it came as a bit of a shock when I realized that Graham Dykstra had been detailed in to take his spot. We hadn't spoken since the phone call over a year ago. There was an air of caution about him as he entered the kitchen and introduced himself to my crew. None of the others had met him, but when my turn came I gave him a polite nod and as friendly a smile as I could muster.

"Hey, Graham," I said. "You know me."

"Hey," was all he said. I couldn't read his face.

Station routine went by uneventfully. He was quiet as we

checked inventory and mopped the floors. There were no calls that morning, and at lunch I had the opportunity to draw him into the conversation.

"So, I hear Captain Peters retired," I remarked. "Did you guys throw a party for him?"

"A small one," he replied. "We took him out for dinner."

"That's nice. How are things at Station 17? I see there've been some changes on the roster."

"Yeah, we lost Kyle. He went to 12. We got a new rookie."

"How about the boat?" I asked. "They were talking about replacing the old one before I left."

"We did eventually," he nodded. "The new one's pretty nice. Powerful motor."

He went on to describe it in detail. The reserve still lingered, but he was warming a little.

At study time, the others scattered, and I was still in the kitchen pouring myself a cup of coffee, when he came back in to retrieve his jacket.

"So, you like it here?" he asked cautiously.

"Yes, I do," I answered.

"Bit of a drive for you."

"True, but it's worth it for a good station."

He wandered over to the window and looked out at the parking lot. I could feel the tension. I edged my way over to the table and sat down. I had had enough of elephants in the room, and it was time to say what had been on my mind.

"I was in a bad place when I wrote you that letter," I began. "And I'm sorry for holding the past over you."

He turned from the window, looked at me for a moment, then took a chair opposite me.

"No need to apologize," he said. "I could tell you were in a bad place, but honestly in the long run it helped me. I didn't take it too kindly at first, but over time I began to see that there were things about my leadership style that needed to change. When you sharpen an axe, you make sparks."

"I think you were frustrated with my mistakes," I offered. "And I know there were some competency issues learning the boat."

"Yeah, there was that," he admitted. "But if I had tried to help you instead of just forcing you along, maybe things would have been different."

"There were a lot of factors at play," I said. "I know you wanted me to be more talkative, but when you're getting treated a certain way, it's hard to be motivated to be part of the crew."

"Looking at it now, I can see that that's what was holding you back," he replied. "I had thought of saying something to Kyle about the way you were getting treated, but I didn't want to make waves. If I could do it again, I definitely would have intervened. If you're not part of the solution, you're part of the problem."

I nodded.

"I want you to know," he went on, "that you never have to worry about things being the way they were again. I've been acting as a lieutenant a few times, and I've learned that people respond better when I'm not so hard on them. Sometimes I've had to give correction, but I always try to do it gently now."

We talked for two hours. At the end, I could sense that a reconciliation had occurred beyond what I had thought possible.

People can change, I thought. And who knows, maybe my small acts of forgiveness were the seeds that had made it possible. I could certainly feel the change in myself.

Something about my work was changing, too. Maybe it

was the daily prayer time, maybe it was what I was reading, but there was a difference that could be felt at the root of things. At first it was barely noticeable, but a sense of peace was growing. Instead of worrying about whether I would make a mistake on the next call and drawing down a reprimand, I would find myself offering up a brief prayer on the way to the call: *Lord, help me to do a good job, for the sake of the public I serve and my crew who are relying on me. If I screw up, I offer that to you. Bring some good out of it.* With these little acts of surrender came a greater sense of freedom, and the tension that used to accompany my reporting for duty was easing. Not only that, but the chest pains and insomnia that had dogged me at Station 6 were now a thing of the past.

It was during this time that my Mom gave me a brown scapular. At first, I was a little reticent to wear it to work, since scapulars have an odd habit of popping out at inconvenient moments and getting tangled, but in time I couldn't imagine leaving home without it. The scapular was a tangible reminder under my uniform shirt that I was consecrated to Jesus through Mary, like my spiritual "dog tags." I pinned a medal of St. Joseph on one end and St. Thérèse on the other. St. Joseph had become a special patron of mine. I looked to him as an example of following Christ in the hidden and the ordinary. My experiences at Station 17 had taught me that I would never be an eloquent evangelist, but perhaps I could make a difference through quiet example. Speaking fondly about my family to crew members, refraining from gossip or conversation that demeaned women, or simply being a good brother to them could be ways of living out the Gospel. St. Thérèse was another example of this. Her "Little Way" of doing simple tasks well out of love was beginning to transform the drudgery of station routine. Cleaning washrooms, mopping floors, and maintaining equipment had become little gems to offer up for special intentions. Over time, I even began offering up the occasional corrections I received for my mistakes. *For the conversion of my co-workers,* I would say; or, *For the end to abortion;* or *In reparation for my sins.* Paul's statement in Colossians 1:24 was starting to take on more and more meaning for me: "Now I rejoice in my sufferings for your sake, and in my flesh I complete what is lacking in Christ's afflictions for the sake of his body, that is, the Church."

At times I still worried whether I would be able to make a good defense of the truth if contentious issues like abortion, euthanasia, or same-sex marriage came up. The personalities at Station 8 made for a very laid-back workplace culture, though, and my co-workers seemed instinctively to keep away from such topics. One day, however, I was given a chance. Clay and Tyler were both off, and we had two firefighters detailed whom I didn't know very well. I chanced to walk into the kitchen as they were in the midst of a discussion.

"My Dad's got dementia," one of them was saying, a sad look on his face. "I just don't know what quality of life he can expect. Me and my sister are going to have to have a conversation about what to do."

"I dunno, Don, that's tough," the other replied. "Like, do you pull the plug just cause he's not quite with it?"

"I'm not saying pull the plug," said Don. "I'm just thinking about it, that's all. He's got some years left. But at what point do you say enough is enough. I know for me if I ever get like that, take me out. I don't want to be a burden to anyone."

Should I jump into their conversation, I wondered? I didn't exactly have an articulate argument against euthanasia. *Come, Holy Spirit,* I prayed. Then I remembered something a pro-life apologist had once said. The most powerful way to change minds and hearts is by asking questions and telling stories. I realized I had a story that could speak to this situation.

"My grandma had dementia," I said quietly. "Kind of like your Dad."

"Oh yeah?" they both looked at me with interest.

"Yeah," I said. "She started coming down with it in her mid-eighties. By the time she was ninety she was nearly stone deaf, and sometimes you had to remind her who you were. I never felt like she was a burden, though. Used to visit her on Saturdays. It was a neat experience. When I started having kids, she got to meet her great grandchildren. It was a really special time. I know she wouldn't have traded it for anything. She lived to be ninety-four."

Don nodded a little, saying nothing in reply.

"That's a good point," his friend said musingly. Their conversation drifted to other things, but I had hope that a seed had been sown.

Thank you, Lord, for the inspiration, I thought.

Boom! I jumped involuntarily and paused my workout, looking out the window. Dark clouds had gathered, and flickers of lightning chased each other through the billows. I walked out into the hallway and met Clay.

"Thunderstorm," he said cheerfully. "That one sounded close."

The door to the truck bay swung open, and Tyler came in.

"I saw a bolt of lightning," he announced. "It looked like it hit real close. Wouldn't be surprised if we get a call."

As if in answer the printer began its rapid clicking, and the flat tone sounded over the speakers. I ran to the watch.

"Engine 8. Odor of smoke inside," I read. "5572 Tay."

My eye scanned the paper and saw that the dispatcher had written an additional comment:

"Caller heard a loud bang and says house may have been struck by lightning."

"Let's go!" I heard Frank shouting in the hall.

Tyler ran to the driver's side and fired the motor, while Clay and I dressed. In less than a minute we were driving down the street in the direction of Tay Avenue. I tightened the straps of my SCBA and lifted my helmet off its hook, ready to don it as soon as we arrived. My fingers felt the small medallion of St. Florian, the patron saint of firefighters, that I kept fastened to the underside of the brim. A Catholic fireman I had known gave it to me when he retired. He had worn it on his coat for thirty years.

Okay, St. Florian, I prayed. *Help me to do a good job.*

I never felt this was superstitious, for all baptized believers are part of the one body of Christ, and those in heaven intercede for us. It was comforting to know that since I was "surrounded by so great a cloud of witnesses," I had a special friend who would go into fires with me.

The address was close to the station, and a moment later we were parked in front. There was no sign of smoke, and the owner, a tall, middle-aged man, was standing outside looking fairly calm.

"Thanks for coming," he said as I walked up to him. "I think lightning might have struck the house. I heard a big bang, and I think I can smell something upstairs."

"All right, we'll check it out," I said reassuringly. Frank and Clay had come up behind me, and together we mounted the steps and entered the front door. The air was clear inside.

"Ben, you head upstairs," said Frank. He and Clay separated to check the rest of the house. I clumped up the curving staircase and stopped on the landing. A bedroom lay to the left. I opened the door and looked inside. Smoke was gushing from every crack, seeping out from behind baseboards and curling up from electrical outlets. I dashed back down the stairs, calling:

"The house is on fire, Captain!"

"Grab a line!" he shouted. Clay rushed out to the truck, heaved the 200-footer out of its bed, and came jogging back in the direction of the house, the hose flaking off his shoulder as he ran. I hurriedly spread out the folds so that it wouldn't kink, then seized the axe and Halligan out of their compartment. Tyler was at the pump panel, ready to send us water. I rejoined Clay and Frank at the front door, and together we donned our facepieces. Clay took up the nozzle. I spun my finger at Tyler, and yelled with all my might through my mask:

"Water!"

He grabbed at one of the levers, and I saw the hose snake and twist as it filled. In a second, the water had reached the nozzle. Clay opened it up to purge the air, snapped it shut, and we advanced into the building. Smoke already hung low in the stairway, but Clay noticed something I had missed. Wisps of vapor were coming from a door frame to the right of us. The door to the basement.

"Basement!" he shouted at Frank, who was already halfway up the stairs. He swung the door open, and a wall of black smoke puffed out at us. The hallway went dark to our waists. We crouched down where we could see. Frank joined us and began giving orders.

"Clay, you and I will go down into the basement and fight the fire. Ben, you stay up here and feed us hose."

They disappeared into the darkness. I felt a tug on the line and began shoving as much hose as I could down the stairs after them. My head swiveled around me as I worked. The smoke was banking down farther and farther, and on the wall behind me I could see a flicker of flame behind a light switch. Sirens sounded in the distance. It was in the walls and all around us.

I waited, sipping my compressed air slowly to conserve the supply. I could hear nothing from the basement. The sound of fire was distinct now, a crackling and popping from inside the wall behind me. Truck doors banged outside, muffled voices sounded, and footsteps thudded in the entrance way. A crew of three appeared through the smoke, their breaths Darth Vader-like through their masks.

"Where is it?" one of them asked.

"Basement!" I answered. "And it's spreading up inside the walls."

There was a noise of feet on the stairs, and Frank appeared alone.

"I think we got it mainly knocked down," he said. "I left my guy down there in case it flares up. Can you three go down and give him a hand?"

"Okay!" They brushed by us and were soon lost to view.

"I think it's in the wall behind us!" I informed Frank. "And above us, too!"

"I'll get you to open up that wall, then," he said. "I'm going to get another crew in here with a line."

He disappeared in the direction of the front door. I wielded my axe and began raining blow after blow at the stretch of wall opposite the basement door. As the drywall fell away, spurts of orange flame leaped out, and, as they did so, the darkness in the room deepened. I began working my way upward, exposing the burning timbers as I went and opening up the ceiling above. How long till that hose line arrived? There they were. Frank was back with another crew. In seconds, a powerful spray was washing down the charred studs, and the fire was dying in an angry hiss and a gush of steam. More crews were arriving now, bustling about us in the dark, performing searches and dousing hot-spots. All of a sudden my bells went off.

"I gotta go, Captain!" I shouted.

"Here, let's both go," he said. We made our way outside. The sunshine seemed bright in contrast to the smoky gloom of the building. Crews, hoses, and stacks of tools congested the front yard, and the street was choked with fire trucks. I saw Tyler at the pump panel. Multiple lines stretched from Engine 8, but he appeared to be master of the situation. I walked up to him.

"Looking good, bud!" I said approvingly. He nodded back at me with a little smile.

"Your first time pumping. Everything going okay?"

"All good," he said.

"Atta boy," I gave him a brief pat on the shoulder, then hurried to catch up with Frank. We switched our bottles, and not long after Clay came out to join us.

"I think this thing's under control," he said. "The weird thing is, I talked to the Ladder 14 crew, who were up on the roof. They couldn't find where the lightning hit. Not a hole, not a mark. Nothing."

"Huh!" said Frank. "If you ask me, it struck the furnace vent and traveled all the way to the basement. It's like it followed all the steel duct work in the walls and burned everywhere they touched. It was a strange fire."

An hour later, we were driving home. A big cleanup lay ahead, exhaustion weighed on every limb, but I felt a great sense of satisfaction.

"Good job today, guys," said Frank.

"Thanks, Captain," I replied.

Clay erupted into an off-key rendition of a pop song, while Frank guffawed at the interruption.

There's no job like this, I thought with a chuckle.

14

SUBSTITUTES

"It's going to need another surgery," Frank held up his arm and looked it over critically.

"Another one?" I asked. "What's going on?"

Clay, Tyler, the captain, and I were standing by Engine 8 one morning in late May.

"Apparently my radius isn't fusing properly. The doctor wants to open it up and try a bone graft."

"How long are you going to be off?" I asked.

"Six, eight weeks probably," he replied. "Maybe more."

"Convenient timing," remarked Clay laconically.

"Good old Dr. Summeroff," Tyler teased, taking his cue from Clay. "Hi, Chief, I'm having 'surgery.'"

Frank snorted and gave him a playful punch on the shoulder.

"No, it's real enough," he said seriously. "You'll have some guest officers rotating through here all summer."

"Substitute teachers," said Clay. "Party at Station 8!"

Tyler and I exchanged glances. It was always a gamble when an acting captain filled in for Frank. Most were competent leaders, but occasionally a bad apple would show up and make our life difficult for a shift or two.

"When's the surgery?" I asked.

"Next week," he said.

At a quarter after six on a cloudy June morning, I slung my kit bag over my shoulder and headed toward the station doors. It was the first of our shifts without Frank. As I was placing my gear on the truck, a strange noise caught my attention.

An alien spaceship? was my first thought. A moment later, a black Tesla turned into the parking lot, the hum of its electric motor scarcely louder than the hiss of its sleek tires on the pavement. A synthetic beat came from inside, mixed with the auto-tuned whine of a contemporary pop song. The Tesla came gliding through the open bay door, up onto the cement floor, and stopped beside Engine 8. The hum died. A door opened noiselessly, and the car's occupant stepped out. He was about forty, with officer's bars on his shoulders and a pair of aviators with neon-green frames on his tanned face. Catching sight of me, he flipped the shades up so that they rested on the spikes of his gelled black hair.

"'Ay, Benny Boy!" he greeted, showing a set of perfectly whitened teeth.

"Hi, Gino," I replied. "You acting here today?"

"Yeah, man."

"New car, I see."

"Way of the future, man. Won't be long before everyone's driving one."

"You think so? I don't know if many people could afford it."

"Why not? This thing only cost me fifty grand. Once they phase out all the oil and gas, the whole world will be going electric. We wanna save the planet, man!"

He whipped out an iPhone, pressed his finger to the screen, and the car doors locked themselves with a click.

"Whole world will be connected with these things," he added, pocketing the phone.

"So, what's on the agenda for today?" I asked, changing the subject.

"Extrication training this morning," he replied. "Then

this afternoon's pretty low-key. I wanna try out this badminton net I've been hearing about."

Badminton was the new wave at Station 8. Someone on A Shift had started it by bringing in a set of rackets one day, and the thing had caught hold. Since there was no way of sinking posts in the bay floor, two buckets had been filled with gravel and a green T-post stuck in each. With the net hung between them and boundary lines marked out on the floor with masking tape, the truck bay made a sizable badminton court. Clay and Tyler had already shown themselves to be highly proficient at the game, and in our evening matches, I would usually get the worst of it. I had a feeling that Gino shared their athletic prowess.

After truck checks, we drove down to the nearest junkyard. An hour went by, and we worked away at the remains of an old Dodge Charger, honing our skills as we popped doors and dismantled the roof. It was growing muggier by the minute, and the inside of my bunker coat was soaked in sweat. All it once, it began to rain.

"That's enough for today, boys," said Gino. "I think we better cut it short and head back."

Rain was coming down in sheets, and the roads were slick as Tyler drove us back. Clay belted out a country song off-key.

As we pulled into the station, however, the downpour had slowed to a gentle drizzle. We parked inside and began stripping off our wet gear. I was just hanging my coat on the door handle to dry, when the tones sounded. Tyler ran for the printout, and in a moment his voice sounded over the speaker:

"MVC! Harvest and Sun Valley. Two vehicles involved. One occupant trapped."

He emerged from the watch and ran for the truck. He

fired the engine while Clay and I donned our gear again and Gino hopped back into the captain's seat. A few seconds later we were rolling.

"It's in 14's district!" I called, checking the cab-mounted GPS. "Go south on 8th Line, then right onto Sun Valley Drive at the lights."

"Got it!" Tyler yelled back. I slipped on a pair of medical gloves, then put a larger pair of mechanic's gloves on over top. Clay was struggling into a green reflective vest. I reached under the seat to grab mine. It fitted awkwardly over my suit, but it would add a level of visibility while working in traffic. My skin felt cold and clammy against the inside of my wet coat.

"Tools or first aid?" Clay asked.

"I'll go tools," I said. He nodded, loosed the straps on the med bag and slid the oxygen out of its compartment, ready for use. I rummaged in the container in front of me and found the tool pouch, visually checking its contents: screwdriver, wire cutter, wrenches, window punch, seat-belt cutter.

Radio reports were floating in:

"Command from Engine 14, we're on location. We have a four-door sedan with side impact, driver is trapped. Extrication required."

I felt the truck rock as Tyler accelerated harder.

"Is Engine 14 a rescue pump?" asked Gino.

"No," Clay replied. "They'll need our extrication tools."

In minutes we were arriving on scene. Engine 14 was parked at an angle to block traffic, and two police officers were directing cars around them. A minivan sat sideways in the oncoming lane, its front end crushed in. Two occupants stood by the side of the road, apparently unharmed. At first I couldn't see the second vehicle, but then I spotted it, tucked into a small alcove between a cement jersey barrier and a power pole, with a foot or two to spare on either side. Behind it lay several rows of vehicles, and I realized that we were near a second-hand car dealership. Looking at the car, I could see that the driver's side was badly crumpled, and an elderly man sat immobile behind the wheel. How he had ended up there without hitting either the barrier or the pole was beyond me, but the only impact seemed to be from the other vehicle. A firefighter from Engine 14 was crouched in the seat behind him, hands carefully reaching around the head rest to hold his neck steady. Other crew members were stabilizing the car, cribbing the wheels to prevent movement and popping the hood to disconnect the battery. Clay's med bag would not be needed. Instead, we hurried to the extrication compartment and began lifting out the heavy tools. Clay fired up the portable engine before carrying it over to the car, and Tyler helped me heft the cutters, spreaders, and hoses. We hurriedly attached the hoses to the power unit, hooked them up to the extrication tools, and a moment later I stood with the spreaders ready in my hands.

"We'll take the driver's door first," Gino instructed me. "And the rear door if we need to."

Tyler, thinking ahead, was passing a folded tarp to the firefighter inside.

"Here," he said. "To cover the patient."

Clay reached in the passenger window to help spread the tarp, talking to the patient all the while. The elderly man groaned a little as they spread the tarp loosely over him, leaving him room for air.

"Just to keep debris off you," Clay said soothingly. "You doing okay?"

"Okay," he grimaced. "Is this going to take long?"

"Not long," Clay replied. "We're going to remove a door.

We'll make some noise right by you, but don't worry. Just hang tight."

He gave me the nod. I opened the jaws just enough to grab the door, then squeezed. The metal crumpled, exposing a gap near the lock. I released, closed the jaws, and inserted the tips into this new opening. Spreading a little to gain purchase, I saw the door begin to peel away from the lock.

Lord, help me to do this right, I prayed. The metal began to tear. I stopped. I could just glimpse the nadir bolt down there, with the locking mechanism grasping it firmly. I closed the tips, inserted them deeper into the gap and began spreading again. The door groaned under the crushing force of the tool, but still the lock held. I screwed up my eyes, half expecting a piece to fly up and hit me.

Come on. Come on, I breathed. Suddenly, there was a jolt, and the latch popped free. Tyler grabbed the door and swung it outward to expose the hinges. I put the spreaders down and reached for the cutters. Two quick snips, and the door came off, sheared at the hinges. I looked around for Gino.

"Good enough," he said approvingly. "We'll get him out this side."

He motioned to two paramedics, who stood nearby with a backboard at the ready. They approached and slid the board onto the seat until it touched the patient's hip.

"Hello, sir," one of them greeted. "Are you doing all right?"

"All right," he replied. "Sore all over."

"Well, we're just going to get this backboard under your bum, okay? Then I'll get you to turn and lie down. Can you handle that?"

"I think so," he said, wincing.

"Okay," the medic nodded at the firefighter in the car.

"Keep control of the head." He turned to me. "Can you help me turn him?"

"You bet."

I crouched down near the seat, and they slowly moved the backboard into position.

"Ready whenever you are," the medic directed. Together, we worked the patient gradually around until his back was to us. I took over the head, and carefully, inch by inch, we lowered him down until he lay on the board.

"Go," said the medic. Four pairs of hands grasped the backboard, and in another minute he was clear of the wreckage and being strapped down on the stretcher.

"Thanks, guys," the medic said.

"No, thank you," Gino replied.

They wheeled him away to the waiting ambulance. Doors thumped, and soon they were speeding off with the siren wailing. We found ourselves standing around the wrecked car with nothing more to do. I breathed out.

"Wow," said Tyler admiringly. "You had that door off in, like, thirty seconds!"

"Well, we worked as a team," I replied, smiling a little. "And what are the chances we'd get a call right after training! I'm just thankful it was an easy one."

We worked to clean up the scene, sweeping away broken glass and placing the sheared door off to one side. Clay refueled the power unit and coiled the hydraulic hoses.

"Whenever you're done with that, we can head back to the station," Gino directed. Suddenly, he paused and stepped closer to the vehicle. Reaching inside, he moved something upward with a "clunk". He stepped back with a rueful expression.

"It was in drive this whole time," he said. "Good thing

the tires were chocked. Eight of us working, and no one noticed."

As we rattled back toward the station, I prayed quietly. *Thank you for helping me get that door off.*

~

Study time was over, and there was no sign of the crew. I hastened down the hallway, poking my head into various rooms. The kitchen and gym were empty, and so was the lounge. No one was in the captain's office, and the watch stood unmanned. Did I miss a call? Then I heard noises coming from the bay floor. A faint *thwack* sounded at intervals, accompanied by what sounded like the squeak of tennis shoes on a slippery floor. I pushed open the double doors and saw a fierce badminton match in progress. Tyler was dashing back and forth, batting the birdie with powerful swipes of his brawny arms, while Gino returned his shots with apparent ease, hardly seeming to shift his position at all. Clay stood nearby, toweling his sweaty brow. The bay doors were open, letting in a summer breeze, and a couple of curious passersby were watching from the sidewalk.

"Who's winning?" I asked.

"Gino just beat me," Clay answered blandly. "You're up next . . . once he's finished whupping Ty."

There was a sound of a birdie hitting the floor.

"15!" cried Tyler, puffing. "Good game."

He and Gino shook hands. Tyler jogged off the court and handed me his racket. I gave it a few experimental swings, then walked over to the net.

"All right," I said. "One round."

It was then that I noticed that Gino was sporting a pair of bright pink jogging shorts. I caught Clay's eye for a moment before he turned away with a suppressed smile. It was a short round. Ten minutes later, I was walking off, beaten by thirteen points.

"All right, who wants to go again?" called Gino.

"I'll go," said Tyler gamely. "I need to get you back."

"Getting hot in here," said Gino. He whipped off his T-shirt. Clay's eyebrow went up a fraction. Tyler, pausing on the court to tie his shoelace, registered surprise for a second before forcing his face into a neutral expression. Shirtless and pink, our officer swung his racket around breezily while he waited.

"Let's get some tunes going," he said enthusiastically. Walking over to a bench, he picked up his iPhone. A moment later, the bay was echoing to the sound of the Backstreet Boys. The people on the sidewalk were definitely getting a show now. I beat a hasty retreat into the watch.

There were no calls that night. In the morning, after a brief chat with the on-going crew, I put my suit and helmet away in the gear room. As I was headed to the parking lot, I heard Gino calling after me.

"So long, Benny!" he sang out jauntily.

"See you, Captain," I replied, turning around. "You back next shift?"

"No, it won't be me. See ya round!"

He disappeared into the depths of the Tesla. The motor hummed, the tires moved soundlessly, and in another second he was zooming away, techno beats fading with him. We had finished our first shift without Frank.

Two mornings later, as I was hanging my coat on the engine, I met our newest substitute. A short, sallow figure was shuffling through the back door. A disheveled mop of gray hair topped a face that sagged lugubriously, and washed-out blue eyes held a glint of sardonic humor.

"Good morning, Captain," I greeted. He mumbled something in reply that could have been "morning."

We checked our SCBAs together in silence. Clay arrived a few minutes later. He took one look at the captain, and burst out:

"Happy Harry!"

The captain wheeled to face him, a smile turning down one side of his mouth. He swore.

"Oh, no, not you again! I should have booked off."

Clay chuckled and began setting up his gear. Tyler came in, climbed into the driver's seat, and began filling out the logbook.

"Hi, Captain," he said cheerfully. Happy Harry appraised him for a moment, then swore again.

"You driving now?" he said. "We're really in trouble."

Tyler gave a good-natured grin. "Missed me, Harry? I bet the new rookie at Station 1 isn't doing near as good a job as me!"

Harry swore for a third time. "He's awful," he said morosely. "Not as bad as you, but awful."

Nice, I thought to myself. *A sarcastic officer.*

Later, as Tyler was whipping up a batch of eggs, Harry banged around the kitchen, muttering under his breath. He disappeared into his office, and I could hear him talking into the phone. He reappeared a moment later and looked the three of us over critically.

"I've ordered you a blender," he announced. "What kind of station doesn't have a stupid blender? Now how am I going to make my morning smoothie?"

"Rough life," remarked Clay.

"There's a hand blender in the drawer," I pointed out. "You're welcome to use it."

"I guess it'll have to do," he grumbled, opening the drawer. Lifting the lid off a lunch cooler, he began dumping frozen blueberries, protein powder and leaves of kale into a large cup measure. Walking over to the fridge, he got out two eggs, a jug of milk, and some ice cubes, which were soon added to the mixture. He wielded the hand blender skeptically. A loud whirring sound filled the kitchen as he inserted it into the mixture. Greenish globs spattered all over the counter and the front of his shirt. He held it down doggedly, cursing all the while. Finally he gave up and tossed the offending tool into the sink.

"For crying out loud," he spat out. "Look at this mess! Rookie!" he barked at Tyler, "Clean that up!"

"You leave our rookie alone!" I shot back at him. "You're just a guest officer!"

He stopped and looked directly at me for the first time, a toothy grin lighting up his face.

"All right," he said. "You gonna be mouthy, we'll just have to do four hours of blacked-out search and rescue drill this afternoon."

"Go for it," I replied. "I'll be doing my own blacked out activity . . . in the dorm."

The grin widened, then grew into a chuckle. "Buncha rebels," he said. "A guy can't get any respect around here."

He wiped up the splatters, scooped up the half-blended concoction and shuffled away, still grinning.

"If anyone wants me, I'm in my office," he called over his shoulder. "Drinking swamp water."

When he was out of earshot, I turned to Tyler.

"Does he bother you?" I asked.

"No," he replied, smiling a little. "I actually really enjoyed working with him at Station 1. He has a very negative sense of humor, but he's a good officer. Always looking out for his guys."

"As long as you're okay," I said.

"Oh, yeah. Also, he's a total Black Cloud. It's awesome."

Firefighters didn't tend to be superstitious, but if someone was dubbed a "Black Cloud", it meant that they got more than their fair share of fires. Not only that, but we could expect other serious calls, like a VSA or an extrication, whenever a Black Cloud was detailed to our station. Why some people had this luck and not others remained a mystery, but after six years on the job, I had to admit that it was a real phenomenon.

"Looks like we're in for a busy day then," I said.

As if on cue, the call came in at 2 P.M.

"Working Fire! 1309 Highland Avenue! 14's District."

Tyler made an excellent show of response driving. Careful, smooth, fast, but not too fast, he navigated the thick afternoon traffic and had us arriving on scene just as Engine 14 was pulling off their first attack line. It was a two-story home, with smoke pouring out the door and puffing from the eaves.

Harry, Clay, and I walked up to the front yard with our hands full of tools, while Tyler ran to help Engine 14's driver hook up to the hydrant. The attack crew dropped to hands and knees, donned their masks, and advanced the hose line, now charged, through the front door. Harry was calm, almost nonchalant, as he gave us our orders. Gone was any trace of hostility or sarcasm.

"Okay, guys," he directed. "Grab a second line off of 14. Once you mask up, we'll be going in to help them."

I pulled 200 feet of hose off the back of the truck, flaked it out over the grass to prevent tangles, and signaled to the driver for water. While it was charging, I dropped down to put on my own facepiece. I turned on my bottle and felt the rush of cold air against my face. Clay and the captain had theirs on as well.

"Clay, take the nozzle," Harry ordered. "Ben, you help push hose."

We entered the dwelling. Smoke was banked to the floor, and we lost visibility instantly. Clay crouched down lower and began following Engine 14's hose line by feel. He turned to the right and disappeared.

"Basement stairs!" I heard him calling.

"Go down!" Harry replied. I heard his boots thumping on the stairs, and I followed, grasping the hose line and my axe. Suddenly, my feet flew out from under me, and I felt myself falling. Only for a second, though, for an instant later I was lying on something hard. I had tumbled down a short flight of stairs and was on some sort of landing.

"You all right?" Harry's voice was anxious behind me.

"I'm fine. Be careful as you come down. I can't see a thing."

"Hold on!" he answered. I scrambled to my feet and crouched, feeling the mounting heat. I heard him key his mic.

"Command, from Engine 8. Fire's in the basement." There was a pause. No answer came over the waves. He tried again.

"*Command from Engine 8! Do you copy?!*"

Still no answer. He spat out a swear word.

"No reception in here. I'll have to go out and give an update face-to-face. Stay here and stick to the hose line. If Engine 14 leaves the basement, come out with them."

"Yes sir," I answered. He vanished.

A second later, a mask loomed through the smoke. I recognized Engine 14's captain.

"Hey!" he shouted, shaking me by the SCBA strap. "There's no comms in the basement. I can't get through to command. Tell them it's an electrical fire, and we can't get water on it until the power's shut off! And hurry! It's getting hot down here!"

I scrambled back in the direction of the front door, following the hose. Before I got there, I bumped into Harry, who was making his way back.

"We gotta shut down the power!" I said urgently. "Tell command! They've got no radios down there, either!"

He turned around and was gone. I could hear radio chatter outside. In a moment he was back.

"Come on," he said.

Together we made our way down the stairs, feeling the temperature intensify as we descended. My ears began to tingle. There was movement and shouts. A crackle of flame sounded from somewhere nearby. Someone bumped into me in the dark. I heard Harry shouting.

"Power company's on scene. They'll have it shut down in a minute. Don't spray yet."

"How'll we know without radios?" came a reply, high pitched and anxious.

"I'll run back up!" he said. I felt him brush by me in the dark. I kept a firm grasp of the hose, waiting. Where was Clay? It was getting unbearably hot.

A muffled shout came from the top of the stairs, somewhere above me in the dark.

"Hit it!"

I felt the line jerk in my hands. There was a swoosh of water and a hiss of steam as two nozzles doused the flames.

The heat lifted a little. Flashlight beams jabbed around me in the smoke.

"We need ventilation!" Engine 14's captain shouted.

"Roger!" I heard Harry's voice. Shortly after, there was a roar of a gas motor. I felt a rush of air around me. Smoke swirled and lifted, revealing the shadowy shapes of firefighters. I spotted Clay. His low-air alarm bells were sounding. Together, we retraced our way up to the main floor. Harry met us at the top of the stairs and beckoned for us to follow. We emerged into the June sunshine, stepping around the high-powered fan as it pumped fresh air into the structure. We pulled off our masks.

"These radios are a piece of crap," was Harry's first comment. "Good job relaying info though," he continued, nodding at me. "That was touch and go in there for a bit."

There was some overhaul to do, but in a short amount of time we were clear to return to station. We gathered up our tools and made our way back to Engine 8. It was then that I noticed Tyler, off to the side, squatting down beside a dog. It was a beautiful animal, a young golden lab, panting happily as he stroked its head. A group of civilians stood around them, a middle-aged couple and a teenage boy and girl. The owners of the house, I presumed.

"Who's this, Ty?" I asked, walking up to them.

"I found this feller in the house," he replied.

"Yes, he saved our Toby," the woman said, wiping back tears. "I'm so grateful."

"Where was he?" I asked.

"Right beside the basement door," answered Tyler. "There was nothing more to do on the street, so I put on my SCBA and came in to give you guys a hand. He was just lying there on his doggie bed, right beside your hose line."

"No kidding!" I marveled. "I must have gone right by him in the dark. But he seems perfectly fine! You'd never know he breathed in all that smoke."

"He was alert," Tyler said. "He just looked at me, let me pick him up."

"Huh!" was all I could say.

As we drove back, Tyler looked over his shoulder at me.

"Told ya," he said.

"Told me what?" I asked, puzzled.

"Black Cloud," he answered, nodding at the captain.

Harry was scheduled to be our officer again the next shift, but we had a five-day stretch off before returning to duty. I was grateful for the break. It had been an intense couple of shifts. At home, I noticed that we had an interesting new magnet on the fridge. It was a miniature blender, perfect in every detail. Something clicked in my mind. I absconded it and stowed it in my kit bag, so that I wouldn't forget to bring it to work next shift.

Harry shuffled in at a quarter after six the following Monday. His greeting was cordial, if a little mumbled, and he seemed to be in a good mood. Once he had gone into the kitchen, I rummaged around in the utility room until I found a large cardboard box. Borrowing a felt pen from the watch, I wrote on the box in the same manner that the maintenance department labeled their deliveries:

"Attn: Capt. Harry Armstrong. Station 8. Item: 1 Blender."

I dropped the miniature inside and taped the flaps as professionally as I could with clear packing tape. Clay, Tyler, and the captain were sitting around the table as I carried it into the kitchen.

"Something for you, Captain," I announced, throwing a little note of excitement into my voice.

His eyes lit up as they fell on the label, which I had been careful to leave facing out.

"Already?" he said. "That was quick!"

"You wanna do the honors?" I asked, placing it on the counter.

"Sure!" he jumped up. He pulled out a small pocket knife and began carefully slicing the tape. He lifted the flaps and peered inside. There was a moment of blank incomprehension, and then a humorous expression began to creep over his face. He lifted the tiny blender out, holding it up between his thumb and forefinger.

"That's a good one," he said, squinting at it and beginning to chuckle. "I'll have to take a picture."

Tyler and I were laughing now, and Clay indulged in a small grin. Harry set it on the counter beside the box and used his cell phone to snap a photo.

"No respect around here," he lamented, cocking his head at me. "I guess I'm just a guest officer after all."

"Distinguished guest," I replied.

It was my turn to drive again. We had one minor call in the afternoon, a false alarm at the strip mall, and the rest of the time was filled in with maintenance, training, and, of course, badminton. Harry showed himself to be adept at the game, beating me 15–7, and narrowly losing to Tyler in a heated round. At ten o'clock, I turned in to my bed in the watch, and in no time at all I was fast asleep.

The station printer startled me awake. I stumbled over to the desk and switched on a light. Tearing off the paper, I held it up, forcing my bleary eyes to focus. The long flat tone was sounding over the speaker. I keyed the mic and read out our assignment:

"Engine 8, MVC on Highway 141, East of Fairmont

Parkway. Head-on collision. One vehicle on fire. Occupants trapped."

I ran for the truck and fired the motor. There was no need to check my route, I knew the way to the highway. A quick glance at the clock showed the time: 3:20 A.M. This sounded like a bad call. The others wasted no time in getting to the bay and throwing on their gear, while I opened the garage door and switched on the truck's emergency lights.

"All set?" I called.

"Ready," said Harry. I released the air brakes and stepped on the accelerator. Engine 8 roared out of the garage and swerved right onto the street. It was dark and empty, except for occasional patches of pavement illuminated by street lamps. Red flashes from our emergency lights reflected on traffic signs. I blinked to clear the sleep from my eyes and scanned the road ahead.

"Engine and Rescue 16 will get there first," said Harry, glancing at the GPS. "We should get an update soon."

I turned off of 18th Avenue onto Town Line Road, heading north toward the freeway. It would be a good five minutes till we arrived.

"Dispatch, Rescue 16 on location with Engine 16," a voice crackled over the waves. "We have one vehicle, fully involved, and another vehicle with one occupant trapped. Engine 16 will be fire attack. Rescue 16 will begin extrication."

My pulse quickened. I stepped harder on the gas. The truck lurched around a bend, and Harry clutched at the dash to steady himself. We topped a ridge, and I could see a traffic circle ahead.

"Slow down," he said. I eased onto the brakes, and we entered the circle, exiting a moment later onto Fairmont

Parkway. A straight run lay ahead, and I brought the truck back up to speed, listening as new reports came in.

"Dispatch, from Highway 141 Command. Fire is knocked down. We've checked the vehicle. No occupants found. Extrication operations are underway on the other vehicle."

The on-ramp for the highway was coming up on the right. I slowed and steered into the access lane. We took the ramp, came around the bend, and saw rows of red and blue flashing lights a good three hundred yards ahead. The accident appeared to have happened right where traffic merged with the freeway, and police cars and fire trucks were lined up in single file along the ramp. I brought the truck to a halt behind Engine 16. We gathered up our tools and first-aid equipment and began walking toward the scene. A police officer met us halfway.

"There's someone missing," he informed us. "A child. Can you guys search the ditch?"

"Drop your stuff," Harry ordered. "Grab your flashlights."

I dropped the med bag and switched on my light. Together, we climbed over the roadside fence, scrambling down the steep slope, feet sliding in the soft gravel. Firefighters from Engine 16 were already at the bottom, pacing back and forth among the bushes and trees and hallooing as they combed the area. We joined in the search, but after several minutes it became clear that no one had been thrown from the vehicle. I followed the captain as he struggled back up the bank, pausing at the top to take in the incident. The smoking remains of a Honda Civic lay against the railing. Another car sat a few yards away in the opposing lane, crushed to almost half its depth. Glass, tires, and other debris littered the roadway. There was a hum of extrication tools as Rescue

16's crew worked urgently to remove the driver. I stuck my head in the window of the burnt vehicle, but there were no bodies to be seen. Whoever the occupants were, they had simply disappeared. There were no bystanders.

I heard footsteps, and another police officer approached us.

"We found the kid," he said, pointing down the highway. "He's in that pickup. But he's in rough shape. Can one of you tend to him until the ambulance arrives?"

"I'll go," I said.

"Right," said Harry. "The rest of you help the rescue crew."

I found the med bag where it had been dropped and made my way at a quick walk in the direction of the pickup. The driver's door opened, and a heavyset man in an orange construction shirt and hunter camo ball cap stepped out.

"I found him on the side of the road," he said, his face white. "He's pretty upset. Doesn't speak much English."

"What about you?" I asked. "Were you involved in the accident, too?"

"No," he said. "I came up on it a few minutes after it happened. I called 911."

I opened the passenger door. A sandy-haired boy with a tear-stained face was sitting there, hunched over and staring at the floor.

"Hey," I said in a friendly voice. "How are you?"

He didn't look up.

"Hey," I said again. "Are you okay?"

He shook his head.

"What's going on?" I asked. "Where are you hurt?"

He turned finally and said something in another language. French, I thought.

"*Parlez-vous anglais?*" I asked, one of my few phrases.

"*Oui,*" he answered faintly.

"Where does it hurt?" I asked again, slower and louder this time. He rubbed his chest. I looked closer. Where his shirt sagged there was an angry red welt. It stretched at an angle to his shoulder and spread upward to his throat, right where his seat belt must have caught him at the moment of impact. He rubbed the back of his neck and started crying again.

"Okay, buddy," I said cheerfully. "Don't worry, I'm going to take care of you. Just stay right where you are."

I opened the rear door and climbed into the crew cab behind him.

"I'm going to put my hands on either side of your head," I told him. "To help you remember not to move. We're going to wait here until the ambulance comes, okay?"

I couldn't tell whether he understood this speech or not, but he made no protest as I held his head. I could hear a siren approaching.

"*Maman,*" he groaned. "*Où est maman?*"

"The ambulance is coming," I said. "They'll take very good care of you. Have you ever had an ambulance ride before?"

He didn't answer.

"How's he doing?" the construction worker asked anxiously, hovering beside the truck.

"Bad whiplash," I answered. "But he's conscious. What was he like when you found him?"

"He was just wandering toward me. I have no idea how he got out of the car. It was totally on fire."

"What about the driver?" I asked.

"Last I saw her, she was running down the road. Drunk, I think. Just left him!"

I pursed my lips in disgust. What kind of life did this child have?

"*Quel âge as-tu?*" I asked him.

"*Huit,*" he answered. Eight years old. It was then I remembered that it was the morning of my daughter's birthday. She was turning eight years old. A lump came into my throat.

"You're going to be okay," I said. The words sounded hollow.

Flashing blue and red lights reflected in the rearview mirrors. There was a thumping of doors. The paramedics had arrived. Footsteps crunched on the gravel, a flashlight shone in, and a young man with a yellow crash helmet and reflective vest looked us over.

"What have we got?" he asked.

"Sore neck and seat-belt rash," I answered. "He was in a head-on collision. He's doing all right, though. They're going to need you guys for the other patient, it's much more urgent. We'll wait here."

"All right," he said. "There's a second ambulance not too far behind us." He and his partner moved off, pushing their stretcher in the direction of the extrication. A second siren could be heard in the distance.

"Not long now," I said encouragingly. "You're doing great."

The volume of flashing light in the mirrors increased. Doors thumped again, and a second pair of medics approached us.

"We got a young patient, I hear," one of them said brusquely, the older of the two.

"Yes," I answered. "Do either of you speak French?"

"I do," the younger one replied. He stood beside the open

passenger door and began addressing the boy in a steady stream of fluent French. He replied in kind, his reticence seemingly gone.

"We'll take over here," the senior medic told me. "Thanks for looking after him. You can release the head for now."

I let go, climbed out, and took one last look at my patient. He was talking animatedly now, more comfortable, no doubt, with someone who spoke his own language. I hurried to find my crew. I saw them standing around the remains of the second vehicle, talking with the firefighters from Station 16. The extrication had been successful, and the paramedics were wheeling a middle-aged male patient back toward the waiting ambulance. He was collared and strapped down tightly. A thick bandage was taped over his forehead and nose, and blood stains streaked his face. I examined the car curiously. Both doors and the roof had been removed, and the dash had been lifted with the hydraulic ram. It had been a serious operation, much more involved than my simple door pop of the other day.

We helped put away the tools, swept debris off the road, and repacked hose.

"What happened when you first got here?" I asked Engine 16's operator, as we reloaded the bumper line together.

"I was busy with the pump," he said. "So I didn't really get to see what was happening, but one of the other guys said they saw the chick from the Honda try to flee the scene. She didn't get far. Cops were coming westbound and picked her up. Last he saw her, she was being handcuffed."

"That poor kid," I said, shaking my head.

"Yeah," he agreed. "I was talking to a cop a minute ago. Apparently she was wanted in a custody case. Wasn't supposed to be anywhere near the boy. Drugs and all that, you know. Somehow she got ahold of him and was trying to leave the city."

Suddenly I realized what a lot I had to be grateful for in my own life. Whatever my problems were, they were nothing compared to what that child had suffered.

Lord, I prayed. *Help this poor family. And help me always to be a good father.*

There were no tears this time, only empathy, and a certainty that I would never forget what I had witnessed on a dark stretch of highway at three in the morning.

~

There was another side to Harry that came out unexpectedly one morning about a month later.

"Who was that on the phone?" he asked casually, as I came out of the watch desk.

"That? Oh, it was Ted English," I answered.

"Our union president!" He sounded surprised. "What did he want?"

I sighed inwardly. I would have to open up a can of worms.

"I'm looking into taking a leave of absence," I replied, then added: "For family reasons."

His face registered interest.

"Everything okay?" he asked.

"Oh, it's my wife," I said. "She's developed some health issues over the last few years. We haven't had much luck getting to the bottom of it."

The sardonic mask was gone now. There was something in his eyes I hadn't seen before, sympathy perhaps.

"That doesn't sound good. It's not cancer is it?"

"No, no. Nothing life-threatening. It's more like chronic

fatigue, you know, and headaches, stuff like that. We've been bouncing around from doctor to doctor, and no one seems to be able to help her. Our best guess is it's Lyme disease."

"Ugh!" he sucked in through his teeth. "Ah, I'm sorry to hear that. That's no fun." He looked thoughtful for a minute. "What did Ted say? Did he think you'd get approved for a leave?"

"It'll depend on a few factors, I think. He said management hasn't been very generous about granting time off for irregular stuff like this. The problem is, it might involve some serious travel. The clinics I've been looking into are a long way away, and treatment can take months."

"That is tough," he ruminated. "Tell you what, I have an idea. If they say no, give me a call. I'll organize guys to work your shifts for you."

I was somewhat taken aback. I hadn't expected such a big offer.

"Really, Captain? That's awful good of you."

"Sure," he said, waving me off. "Don't worry about it. It's not the first time. Remember Jason Martins, the guy on A Shift that died of cancer last year? He had about a hundred people lined up to cover his shifts while he went for treatment."

"I know, but that was a line of duty death," I protested. "I can see people stepping up then, but this is a little different."

"It's not that different," he said. "And he's not the only one. There was Jack DeSippio a couple years back. His kid had major surgery. Guys worked for him for two months. And Craig What's-his-name on B. His wife had cancer. He was off for a year or so."

"That's true," I admitted.

"Think it over," he said. "I'm not saying you have to, I'm just saying it's an option. I'd be happy to organize it."

"Thanks, boss, I will, for sure."

Whatever else he would have added to the discussion was left unsaid, for at that moment we heard the printer clicking down the hallway.

"There's a call!" I said, jumping up.

We pulled up on an abandoned elementary school building issuing smoke from every crack.

"Shit!" spat Harry. "That fire could be anywhere! Keep a sharp eye, boys. There could be squatters in there!"

I grabbed the tools and met Tyler in front of the main entrance as he dragged up the 400-footer.

Good lad, I thought. He had made the right choice for a building this size. Who knew how far into the interior we would have to search before we found the fire. Behind the double glass doors I could see that the smoke was already banked to the floor.

"Take the door," ordered Harry.

I swung the long-handled sledge hammer with all my might, and a pane of tempered glass disintegrated in a musical shower. Another swing annihilated the second pane, and a quick stroke broke off the panic bar that stood at waist height. Smoke gushed out. We donned our masks quickly while Clay charged the line, then we were advancing into the blackened depths of the unfamiliar building. It was hot. I felt the oppressing vapor pushing down on us, almost like a physical weight. We were bearing slightly left from the doors. Every so often, we would stop to feel our way around some obstacle. It was impossible in the pitch black to know what we were bumping into. Shelves, I thought.

"It's hotter to the right," I called out.

"What?" came Harry's voice, oddly muffled as if he were

a long way away, though he couldn't have been more than six feet in front of me.

"Fire's to the right!" I yelled.

"It's above us!" Tyler's voice.

"We're too far this way," I thought I heard Harry say. I felt the line loop around my torso. They had switched directions. I turned with them, and the line straightened. Which way now? I kept a firm grasp on the hose and felt my way along it. Every near miss and line of duty death report I had read had started with a firefighter leaving the hose line. I bumped into someone. There was a flashlight beam in front of me.

"Tyler?" I called out.

"This is Engine 16," came the reply. "Who's this?"

"Engine 8!" I said. "Where's my crew?"

"They went right!" he said. "We just passed them."

Right! If they had turned around toward the doors, then they were still moving away from the fire. And the heat was intensifying. The firefighter from Engine 16 had disappeared. Two more shapes bumped into me, heading the way I had come from. I stepped to the right, in the direction Harry and Tyler must have gone. Suddenly, my hand came in contact with another hose. I felt over the top of it to grasp at mine and felt another one. I realized with a sinking feeling that there was a tangle of charged hoses all around me. Engine 16 had crossed our path and brought their own line with them. Somehow, in changing directions, we had looped our line at least once around theirs. There were butterflies in my stomach. I would need to sort out this tangle if I wanted to find my crew. And my air supply wouldn't last forever. I gave a strong tug.

"Stop pulling on the line!" The voice was muffled, but it was Harry. I hurried in the direction of his voice and in seconds was free of the tangle, grasping my own hose line. I saw flashlight beams ahead.

"Engine 8?" I yelled.

"Right here!" Tyler's voice was cheerful. I got close and grasped his SCBA strap.

"Still no sign of it," said Harry. "We better get out while we still have air."

We retraced our steps, holding on to the line for dear life. When we got to the tangle, some movement of air had lifted the smoke a little, and I could spot the open door a few yards away, a dim patch of blue light through the haze. We hurried toward it and emerged into the sunlight, just as my bells went off.

There was a roar, and a window failed a hundred feet to the left of us. Flames licked upward, scorching the bricks above it.

"There's the fire," snorted Harry. "For crying out loud. We were searching in the wrong part of the building. Quick, guys. Fresh bottles, then we're bringing in another line."

Clay, like a good operator, had stretched a second line to the door, and in a matter of minutes we were back inside, this time bearing right in some kind of hallway. Harry walked ahead, invisible in the darkness; I held the nozzle, and Tyler took up the rear. Suddenly, we stopped short.

"Line's stuck!" I called. I pulled with all my might, and I could hear Tyler's sucking breath as he exerted himself. The line would not budge.

"I'm going back!" he shouted. "See if I can free it!"

I waited, then began tugging. It moved a few inches, then stopped short again.

"You coming?" Harry had reappeared.

"Just give us a minute," I said. "It's snagged on something."

He disappeared again. I pulled for all I was worth. The line crept forward a few more inches, then stuck. What was going on? I braced my legs, grasped firmly, and heaved, head down, in the direction of the fire. The line gave, then slid, and I was moving forward in a headlong charge. My momentum was my undoing, however, for just then the line caught again, and I crashed forward on my face. My breath coming in gasps, I struggled to my feet. I fumbled around in the dark and found the nozzle.

"You're almost there," Harry's voice sounded ahead of me.

My arms were aching as I grasped the hose again.

St. Joseph, I breathed. *Help me get this line to the fire.*

Summoning all my strength, I gave a gigantic heave. This time the line traveled about ten feet. I could see a doorway to my right and the flicker of flames.

It didn't take long to extinguish the fire, and for the next hour we were busy searching the cavernous expanse for bodies. There were signs that people had been living in it. I could see piles of discarded clothing, sleeping bags, liquor bottles, and other cast-off items, but we found no victims. It was clear that the fire was arson. It had been started in two places, one in the classroom we had struggled to get to, and the other in the gym, which turned out to be not far from where we had turned around the first time. Engine 16 had doused that one. What the motive was for starting the fires we would never know.

"That one nearly had me beat!" exclaimed Tyler, as we drove home. "I thought we weren't going to get down that hallway."

"I was pretty bagged, too," I replied. "Four hundred feet get pretty heavy over that distance, even for a monster like you," I added with a grin. "And once it gets snagged, good luck."

"That hallway was choked with all kinds of stuff," he replied. "It was a total fire trap."

"Let me know next time if you're at your limit," said Harry seriously, turning around. "There's no shame in needing to back out. We can regroup and attack it again when you're ready."

I was hoping for a quiet night after the day's efforts, but just after midnight the call came in for another structure fire. We arrived just in time to see the roof collapsing and 50-foot flames shooting into the night sky. Many trucks were on scene ahead of us, and several dozen firefighters surrounded the house, dousing it from outside. It was a traditional "surround and drown" fire, the kind I often used to fight as a volunteer. Since the structure was fully involved, and no one could possibly survive with that volume of heat, there was no need to risk any crews by sending them in to do a search. Harry directed me to pick up a nozzle, and soon I was lobbing a fog stream in through the living-room window. My shoulders, arms, neck, and thighs still ached from the morning, but I was in my comfort zone now. No getting lost here; this was a fire I could see.

We returned to station as the first rays of the morning sun were tinting the roofs of Lower East Side. It had been Harry's last shift at Station 8, and Frank would be back the following Monday.

"Give me a call," were Harry's parting words as he walked out to his vehicle.

"I will," I said. I watched him go. Tyler was right. He certainly did look out for his guys, and there was no doubt that he was a Black Cloud. I wondered if he was taking all the excitement away with him.

15

PRIDE

When Frank walked in, I greeted him with an enthusiastic handshake.

"Good to have you back," I said warmly. "How did everything go?"

"Went well!" he responded with his expansive smile. "Arm's healing up. It was nice to have the time off, but boy I'm ready to be back to work!"

"Got antsy the last few weeks, huh?" I grinned. "Well, we certainly had enough excitement around here while you were away. Maybe we'll keep our busy streak going."

"Hopefully!" he laughed. "Yes, I heard that Happy Harry brought his usual luck with him. I—" he stopped, staring at the front of the rig. "What is *that*?" he exclaimed.

I followed his gaze and saw that there was a large decal affixed to the windshield. It had not been there the shift before. It bore the usual fire department logo, a Maltese cross flanked by axe and ladder, but instead of the usual navy blue background there were rainbow stripes. A strange feeling came over me, almost like a shiver. It had come to this at last.

"Why is that on there?" Frank demanded loudly. At that moment, footsteps sounded, and the B Shift Captain came around the corner of the truck. He must have overheard the last comment, for he eyed Frank coldly.

"It's Pride Month," he announced. "Every fire truck in the city's got one." *Duh,* his tone implied.

Frank didn't answer, but his brows were knit. The other captain gathered up his bunker gear and headed for the storage room.

"Frank," I said slowly, keeping my voice low. "Could we *not* have that on there?"

His glance was expressionless, but after a short silence he began nodding slightly, pursing his lips.

"If it disappears today," he said, "I won't say anything."

"Right," I said simply. Then a thought struck me. "It's not that I mean any disrespect," I explained. "It's just that I didn't sign up to promote special interest groups."

"I know," he said. "I know. This whole gay pride thing never made any sense to me. People can do whatever they want in their personal lives, as far as I'm concerned, just don't force me to advertise for them. I mean, what's next? Are we going to be putting Greenpeace stickers all over our trucks? Animal Rights? It's a fire truck, for crying out loud!"

This was an unusually political speech for Frank, and I could tell that the matter had upset him. Upper management had supported Pride Month for years, but employees had only ever been involved on a voluntary basis. Usually, a spare fire truck would be selected to drive in the parade, manned by off-duty members. People like myself who had religious objections need not participate. This was the first time the chief had gone so far as to expect on-duty crews to be involved. Now, simply by virtue of showing up for work, we were all being forced to promote the agenda, whether we liked it or not. And we had been given no advance notice.

The man door at the back of the bay clicked, and Clay walked in whistling.

"Morning, Clay," I greeted, a little distractedly. Then I

noticed what he was wearing. It was a standard navy-blue T-shirt, but instead of the normal fire department insignia on the upper right-hand corner of the chest, it bore the rainbow logo.

Oh, no, I thought. Clay was married, with two teenage daughters, but apparently he was fully on board with the cultural narrative. So, it seemed, was our clothing department.

Frank said nothing about the shirt, though I noticed his eye fell on it, and Tyler showed up in a normal uniform. I waited until morning checks were over, and once everyone had gone to the kitchen, I peeled the sticker off the truck. I felt a little uneasy as I tossed it into the trash can. It was a small action, but it could have big repercussions, if I were to be observed by the wrong people. How could I explain that my motives were not those of prejudice or hatred? Yet inevitably they would be construed that way. My mind went back to a conversation at Station 6, over a year ago. Bryce and I had been preparing supper, and Mike McCannell had stood nearby, sharing a piece of news.

"You heard the latest about Doug Morrison?" he was saying.

"The homophobe?" inquired Bryce with an amused sneer.

"Yeah," said Mike. "They just reviewed his case. Looks like they're not going to fire him after all. They'll give him one more chance. Apparently, management told him it would be in his best interests to march in the Pride parade this year."

"What did he do to deserve that?" I asked.

"Oh, he made an insulting remark to some gay guy on Facebook," answered Mike coarsely. "I guess the guy noticed that Doug had photos of himself in uniform. He complained to the city, and Doug got hauled up to headquarters for a nice little chat with the chief."

"Well, he's a dumbass to put anything to do with his job on the internet," said Bryce unfeelingly.

"Yup," agreed Mike. "You have to be so careful these days."

"Wow," I ventured. "I mean, I might disagree with someone's life-style, but I would never say something insulting, especially on the internet. Still, if the chief told me I had to be part of the parade, I would have to refuse."

"That wouldn't go so well," said Bryce meaningfully.

"It would be a career-killing move," added Mike.

"I know," I said. "But don't you think it's wrong to force people to go against their beliefs?"

"Unfortunately, some beliefs are given more weight than others," observed Mike with a snort.

"That's the other thing," I said indignantly. "We're wearing a uniform. It's supposed to be an emblem of neutrality. When I show up at someone's house in an emergency, I don't represent any political cause or special interest group, and people trust me because of that. Now all of a sudden I'm being forced to take sides!"

"I agree with you there," said Bryce coolly. "But I'll just say . . . be careful who you air your opinions around."

"Well, that's not right, either," I said.

"No, but it's the day and age we live in."

Now, I wondered what the chief would think about what I had just done. If a Facebook comment had almost got someone fired, what would removing a Pride sticker do? I thought of what I would say if I were ever called up to the office. 1 Peter 3:15 came to mind:

"Always be prepared to make a defense to any one who calls you to account for the hope that is in you, yet do it with gentleness and reverence."

It was a tall order. How exactly would I express what I believed about homosexuality without coming across as a bigot? I could sense instinctively that something was wrong about being made to promote such an agenda, but I realized that I needed to articulate it for myself before I could make sense of it for others. So, if my opposition wasn't out of prejudice, then, what was it based on? Was it merely the presumed self-righteousness of the moral high ground, a looking down on those who made "inferior" choices?

Lord, I prayed. *Something is very wrong here, and I don't like what is happening. But if you want me to fight it, don't let it be out of pride.*

Engine 8 responded to a call for alarms at Home Depot that morning. Engine, Ladder, and Rescue 16 were on location, along with the District Chief's car, and I noticed that ours was the only vehicle without a rainbow logo. As we were leaving the scene, the chief glanced at our windshield briefly, but whether he noticed the sticker's absence or not I couldn't tell. On our way back to the station, we passed an elementary school. It was recess time, and over a hundred children were out on the playground. A rainbow flag was slowly flapping in the breeze. Written in large letters on the brick wall was the school's name: St. Thérèse of Lisieux Catholic Elementary.

Back at the station, I checked my city email. My inbox held several messages. One was from the Chief of Department and had been sent to every firefighter:

> In our effort to support our brothers and sisters in the LGBTQ2 Community, the fire department will be making use of visual reminders. The city wishes to show the public that we welcome people of all genders, and support sexual diversity in the workforce. This year, each vehicle will display a specially-designed emblem. These emblems must be fixed in a conspicuous location on the front windshield of every engine, ladder and rescue. Chief's vehicles and staff cars will display the sticker on the rear window or

bumper. In addition to these measures, we will be holding a flag-raising ceremony today at headquarters at 1400 hrs. The Pride Flag will then be hoisted each morning for the duration of Pride Month. The chief's office would like to thank the human rights committee for their hard work in designing the windshield logo, and for all their suggestions for this year's Pride Event. Together with our city hall staff and everyone here at fire headquarters, I would like to wish you a very happy Pride Month.

The next email was from the firefighters' union, and was signed by the chairman of the charities committee:

Happy Pride! On behalf of all your elected union representatives, I would like to be the first to wish you a very festive Pride Month! The union takes this opportunity to remind all its members that diversity is one of our core principles, and that we are working hard to create a more safe, inclusive work environment for everyone. Please do your part in making this year's Pride Event the best ever! We are still looking for volunteers for the parade. If you haven't already, please sign up on the union website.

Members are also reminded to wear their Pride T-Shirts while on duty. If you have any suggestions about how to improve this year's event, please do not hesitate to reach out to me. Stay safe, and Happy Pride!!!

I had never seen such enthusiasm thrown into an official email. The union was involved in several other parades and charitable activities throughout the year, but none of them were announced with such gusto. Why the effort, I wondered? Were they compensating for something? And what exactly were we supposed to be celebrating?

During my prayer time, I had a little more leisure to reflect on the issue. I remembered the saying that the safest place for persons experiencing same-sex attraction was in the arms of mother church. There was much to this statement, and it would need some unpacking if I were to use it as my starting point in a defense. What hope did the Catholic Church hold out to people who identified themselves as homosexual? And how was this hope better than what the secular world had to offer? At face value, the Church's call to chastity could be mistaken for a denial of love, and, indeed, this was how it was often portrayed in the media. Yet these same teachings had power to lead people trapped in brokenness to freedom and true love. Years before, I had stood in the rain with 800,000 other young people, as an eighty-two-year old man with Parkinson's spoke to us about this yearning of the human heart.

"The spirit of the world offers many false illusions and parodies of happiness," he had said in his thick Polish accent. "There is perhaps no darkness deeper than the darkness that enters young people's souls when false prophets extinguish in them the light of faith and hope and love. The greatest deception, and the deepest source of unhappiness, is the illusion of finding life by excluding God, of finding freedom by excluding moral truths and personal responsibility."

These words, spoken so long ago by Pope John Paul II, now seemed to me to be prophetic. Had he foreseen the crisis of sexuality and identity that would emerge in my own time? We had reached the point, it seemed, where we were ready to exercise our independence from God in every corner of society, even fire stations.

At supper, the TV was on, and a car commercial flashed across the screen, briefly featuring a same-sex couple happily purchasing a shiny new vehicle.

"Look at that!" said Frank loudly. "Why do they have to show that all the time now?"

"Show what?" asked Clay.

"Gay couples!" he said, his voice still raised. "I mean, I don't care. Do what you like in your spare time, but don't ram it down our throats!"

"They're just trying to gain more acceptance," said Clay mildly.

Frank was still agitated. "Well, I get that they used to be persecuted and everything," he answered, "But it's gone so far the other way now."

"It seems very agenda-driven," I put in.

"An agenda! Exactly! I had this problem at my daughter's high school," Frank went on. "They had some LGBXYZ—whatever the hell they call them now—person come in and give a talk. Jess didn't want to go, and they told her it was mandatory. *Mandatory!* When she dug her heels in, they made her stand in the hall, for punishment!"

"Really?" I said.

"Yeah! Oh, boy, was I mad! I went in there and gave the principal a piece of my mind. I said, 'if you ever do that to her again, I'm going to take you to court!'"

"What did he say?"

"He said, 'You know I'm a proud gay man, right?' I said, 'I don't care what you do at home. Don't push it on my family!' They left her alone after that."

I couldn't help but admire his courage, but at the same time it would never do for me to resort to an angry outburst. His feelings were natural enough, but there was something missing from his approach. Charity, I thought. My position would have to be expressed more carefully. Clay had been concentrating on his plate during the last part of this conversation. He and Frank had worked together for years, and he could probably tell that there was no point getting into an argument. Something about the glint in his eye, however, suggested that the Pride shirt he was wearing might have been a deliberate dig at Frank. Tyler, meanwhile, had been following the story with interest and seemed keen to jump in.

"What about your son's school?" he asked. "Do they do that kind of thing there, too?"

"It's just as bad there," Frank answered. "He's only in Grade 6, but some of his buddies have already started talking about how they think they might be gay. This was after a guest speaker came in, of course."

"They're indoctrinating the children so young," I said, shaking my head. "I hear they introduce the topic in Grade 2 now."

"It's true," said Frank. "And it's not like it would ever cross these kids' minds if their teachers hadn't put it there. I guess that must be why you and your wife homeschool."

"It's one reason, for sure," I said.

"There's a girl in my son's class who identifies as a cat," Frank informed me.

"Cat? Like, short for Catherine?" I asked.

"No, no, she identifies as *a* cat. The animal. The teachers won't let any of the kids call her by her real name. They have to call her 'Cat.'"

"You're kidding!" Tyler's mouth hung open.

"No, I'm not. It's their policy now. If someone identifies as anything, doesn't matter what it is, everyone has to go along with it. They even have a *litter box* in the washroom!"

"No!"

"I'm tellin' you!"

A thought was growing in my mind. At first I hesitated to say it aloud, but after a moment I said:

"See, this is what bugs me. It's all the labeling. Since when did we start putting people in boxes?"

"What do you mean?" asked Frank.

"Well, I'm just thinking back to when I was a teenager. I had some thoughts that I guess you could say were homosexual. I think a lot of teenagers do. Don't get me wrong, I still liked girls and everything, but I was worried about it, thought it meant I might end up gay. Well, I talked to my Dad, and I'll never forget his advice."

I paused.

"What did he say?" asked Tyler.

"He said, 'These thoughts don't define you.' It was really simple, but it totally freed me up. I stopped worrying about the thoughts, and eventually they went away. I guess my point is, people go through stages, and we don't have to label them as gay or whatever just because they're struggling with something. I mean, what would have happened if I'd opened up to one of these modern teachers. They probably would have told me, 'Congratulations, you're part of the LGBT community now.' How confusing would that be for a kid?"

"This is getting deep," said Tyler, grinning.

"Solving the problems of the world, buddy," laughed Frank. "It's what we do in the fire department. But yeah, back to your point, Ben, if I got a label for every weird thought I've had over my lifetime, I'd belong to a lot of different groups."

"Nowadays," Clay offered dryly, "your Dad might even have ended up with a hate-speech charge for that little piece of advice."

"More than likely," I replied, with a sniff.

I thought about the conversation on my long commute home. It was true that many children were being indoctrinated into choosing something that would not have come naturally to them, but what about the people who had no choice? What about those who, through no fault of their own, felt no attraction for the opposite sex? The causes were complex, I knew, and the tragedies of abuse, broken homes, absent fathers, and other childhood factors certainly played a part. Our society had told these hurting souls that they could still find love in this world, and yet the "love" that was being held out as the solution had a deadly twist to it. It was this twist that fire management and society at large were forcing us all to celebrate. How, then, was I to oppose the lies and half-truths hidden in the Pride agenda without losing sight of the compassion and love that people with same-sex attraction needed so desperately? It was time to get some insight into the question.

I rummaged around in the center console, until I found the CD I was looking for. Lately, I was in the habit of listening to audio presentations by various Catholic speakers on the hour-plus drive. I had found these talks very inspiring, and they had greatly helped me in my own faith journey. One talk, in particular, was by a well-known university chaplain and dealt with the issue of homosexuality. I inserted it into the CD player and settled back in my seat, as the city landscape slowly gave way to open countryside. As I listened, I felt that I was absorbing a fresh perspective. The perennial teachings of the Church on sex and marriage were being presented in a way that was sensitive and truthful at the same time. The priest explained how all persons, regardless of their state in life, are called to respect their sexuality and that of others. God has a beautiful plan for sex and marriage, and like all good things, it is possible to misuse and abuse this plan. Homosexual acts, he explained, violate the nature and purpose of sex and so are objectively sinful. However, experiencing same-sex attraction is in itself not sinful; in fact, it calls for great compassion on our part.

People suffering this particular cross can find fulfillment in chaste friendships and still experience authentic love in their lives. The Church never condemns any group of people as particularly evil but, rather, invites us to reflect on how we are all broken and in need of redemption. Father recalled the stirring words of Pope John Paul II: "We are not the sum of our weaknesses and failings, but we are the sum of the Father's love for us." The talk ended with a message of hope for people struggling with same-sex attraction to find healing and fulfillment in Jesus Christ.

Now I understood why I could never drive a truck with a rainbow sticker. It was not that I looked down on a group of people whom I considered "evil". Rather, it was because I could not bring myself to promote a lie that would rob them of the chance of true happiness.

~

"Are you in trouble?" The voice of my union representative sounded concerned over the phone.

"No-o-o," I answered slowly. "I just need a little advice, that's all."

"Okay, go ahead."

"It's about these Pride logos," I began. "You know, on the trucks."

There was a chuckle on the other end. "I knew you were religious. Does it bother you?"

"Well, yeah," I admitted. "I'd say it's a real conflict for me. I feel like I'm being forced to promote something that goes against my beliefs."

He sucked in his breath. "Uh . . . I would say, all it means is that management is just telling the LGBT community, 'you're welcome in the workplace.' No one's forcing anyone to promote it."

"Well that's just the problem," I replied. "Just by driving the truck around, I'm waving that flag. It was one thing when employees could volunteer for the parade, but now I'm being forced to be part of it every time I show up for work."

"How would you feel about working with a gay firefighter?" The question was asked with a hint of sternness.

"I want to be very clear," I answered slowly and firmly. "I would treat *any* co-worker, regardless of his sexual orientation, with equal respect and professionalism. And that goes for members of the public, too. But there's a difference in getting along with someone and being forced to agree with and promote his life-style."

"I see," he said thoughtfully. "So what are you hoping to do?"

"I was thinking of writing to management."

"I wouldn't do that," he said quickly. "It would definitely hurt your career. Tell you what I'll do, though. Let me run this by our human rights rep and get back to you."

"All right," I agreed.

The answer wasn't long in coming. Two days later the phone rang.

"I talked to Janet, the chair of our human rights committee," he began.

"And?" I said eagerly.

"So, I didn't know this, but apparently there's a thing called 'competing rights.'"

"What's that?" I asked.

"Basically, it means that when two kinds of human rights are at odds, one kind usually wins. So, in cases like yours, whenever religious rights come in conflict with sexual rights, whether that's in court or in labor negotiations, sexual rights win, every time."

"Really," I said dully.

"Yes, so she was very adamant that you could never be successful in any kind of appeal against management. Not on this issue."

"It's disappointing, but in a way I'm not surprised," I said.

"No," he agreed. "It's the way of the world right now. And that's not all. I had to put in a good word for you. Janet went off on a bit of a rant when I told her we had a religious member who was opposed to the LGBT agenda."

"What did she say?"

"Oh, she tried to grill me about you, wanted to know if you'd refuse to treat an LGBT patient, said you'd definitely be a threat in the workplace."

"That's ridiculous!" I protested. "When have I ever done anything like that?"

"Well, she doesn't know you, and I made sure she understood that you were a respectful kind of guy. The problem is, these human rights people are going to workshops that definitely put religion in a bad light. They're being educated to view people like you as haters."

"Has she ever bothered to do her research and find out for herself what we have to say about it?" I asked.

"Probably not," he chuckled. "I guess the thing for you to do now is talk to your officer about how to move forward. Whatever solution you come up with will have to be limited to your own station."

"I will," I said, suddenly grateful that I had Frank as an ally. I did not volunteer the information that the sticker had already been removed from Engine 8.

"I appreciate your being willing to listen," I said. "And for putting in a good word for me."

I half-expected the sword of vengeance to drop at any time, but as the weeks passed, no mention was made of the missing sticker. Pride Month came to an end, and as we responded to calls with neighboring stations, I noticed that the logos were disappearing one by one from other trucks, too. Management had succeeded in running a heavy-handed propaganda campaign, but it was heartening to know that there were other souls out there who were not completely asleep.

16

FURY

I woke in the pre-dawn stillness, not sure what had disturbed me. Kate was still asleep. I glanced at the clock: 4:45 A.M., fifteen minutes before my alarm was due to go off. Then I realized what had woken me. Flashes of light were coming through the bedroom window. They flickered, then faded, then started up again a moment later.

Something wrong with the porch light? I wondered. I slid out of bed as quietly as I could and peered out the window. All was dark. The summer night was sultry, and despite the fan that Kate had set up on the bedside table, my pajama shirt was damp with sweat. Making as little noise as possible, I got into my uniform. My car keys clinked a little as I put them in my pocket, and she stirred.

"Going?" she asked in a sleepy voice.

"I'm on my way," I whispered, so as not to wake the children. "Have a wonderful day."

I leaned over and gave her a kiss while my thumb traced the sign of the cross on her forehead. She blessed me in turn. It was our morning ritual.

"Drive safe," she said.

"I always do," I replied, smiling. "And don't forget, you and the girls are coming into the city this afternoon, too."

"I know, I haven't forgotten. We'll all be looking forward to seeing you at my sister's place . . . as soon as you get off work."

On the driveway, I paused and looked up at the sky. The

first glimmer of dawn was showing in the east, but something else was happening that arrested my attention. It wasn't the porch light at all. Racks of towering cumuli were piling up, illuminated every few seconds by dancing flashes of lightning. It was completely soundless, but each flash was startling in its brilliance and intensity. The air hung heavy and hot around me.

I watched the clouds as I drove. As the daylight increased, the flashes grew less intense, and though a few raindrops spattered the windshield, the storm seemed reluctant to burst. By the time I was pulling into the station parking lot, the clouds had dissipated altogether, and a flawless blue dome above held the promise of a beautiful summer day.

"We're scheduled for firefighter survival training today," Frank announced at the breakfast table.

"What module are we on now?" asked Tyler.

"Floor collapse," he replied. "We'll be at Station 16 in the training tower. Today they want us to do the Nance Drill."

"I haven't done that since boot camp," I said. "It would be a good refresher."

"Well you know what they say," said Frank. "Train like your life depends on it, because it does."

"Which one's the Nance Drill again?" asked Clay.

"That's the one where the firefighter's fallen through the floor," Frank explained. "Four rescuers lower a hose line down to him, pull him up."

"Oh, no," groaned Clay. "We don't have to pull *you* up through the floor do we?" He sized up Frank's powerful, 250-pound frame disparagingly. Frank looked offended.

"A trim young gentleman like me, in the prime of my life?" he said stoutly. "I should think so!"

"Who was Nance?" asked Tyler.

"Some firefighter who fell through a floor," I answered. "I guess it didn't go so well for him. Never get a drill named after you."

"It's going to be hot today," said Frank, steering the conversation back to the matter at hand. "We'll get the training over with this morning. If you guys want to dress down a bit this afternoon, and just wear shorts and T-shirt, that's fine by me."

The training tower at Station 16 was four stories high and made of solid concrete. The rooms had been designed so that firefighters could work on their basic skills. There was a plywood maze, with tangles of ropes and wires built in, for practicing escape from a ceiling collapse. There were standpipes on each floor for high rise tactics, and the third floor had a room full of furniture for blacked-out search and rescue drills. Today we were to practice on the top floor, where a special trap door had been made, to simulate a floor collapse. We trudged up the stairs in full bunker gear and SCBA, our footsteps echoing against the cement walls. When we arrived, six firefighters from Station 16 were waiting with a charged hose line, ready to start the day's drill. There were nods of greeting and a few brief handshakes.

"Ready to sweat?" said Frank loudly.

"I'm sweating already," one of them answered. "They're calling for 100% humidity today."

"Well, let's get this over with," Frank replied. "Then I can get back to my nice, air-conditioned station."

Firefighters took turns clattering down the stairs to the room below, waiting until the crew above had lowered a loop of hose down to them. Bracing their feet on the floor, the four rescuers would then haul upward with all their might, pulling the firefighter up through the hole, as he stood on the loop of hose. When my turn came, I made my way downstairs, switched on my air, and craned my neck upward to see the hole.

"You ready?" someone called.

"Lower away," I replied.

When the loop had been dropped so that it was level with my knees, I placed one foot firmly on it, grasped the hose with gloved hands, and shouted:

"Okay!"

"Practice putting in your Mayday," I heard Frank say.

My hand went to the mic of my portable radio, but I did not actually press the "talk" button.

"Mayday! Mayday! Mayday!" I said urgently. "Third floor! Engine 8 Crew! I've fallen through the fourth floor, and I'm running out of air! Send RIT!"

"Perfect," said Frank. "Mayday received. RIT has arrived. They've sent down a hose line for you. Now we'll practice the rescue. Go ahead, guys."

They pulled, and I felt myself sailing upward. There were grunts of exertion from the four rescuers, as my head came level with the trap door. Letting go of the hose, I reached for the ledge, got a knee over, and scrambled to safety.

"Pretty smooth," said Frank. "Of course it's easy with a nice solid floor and perfect visibility. Not so much fun when you're in a panic and the hole keeps giving way every time you try to climb out of it."

"And it's pitch black," added Clay.

"All right," said Frank breezily. "My turn."

His footsteps clumped heavily as he descended, and Clay and I looked at each other ruefully.

"Think we need a couple extra bodies?" he asked.

"Might be a good idea. How much do you think he weighs with all his gear on?"

"Over three hundred, I'd say."

We heard him shouting from below:

"Okay!"

The four rescuers lowered the hose and got into position. Clay and I crouched near the hole, ready to lend a hand.

"One, two, *three!*"

There was a tremendous grunting and heaving, but no Frank appeared. I could hear his one free boot scraping the wall, as he fought to get a grip and push himself up.

"Higher!" he was shouting. "I can't reach!"

Clay and I grabbed the hose and helped the others. Frank's helmet appeared above the edge.

"Pin the line, don't let it slip back!" someone yelled. Two rescuers knelt on the hose, while I lay flat on my stomach and reached down to grab Frank's SCBA strap.

"One more pull, boys!" said Clay. "One, two, three, *Go!*"

With a last straining effort, we hauled Frank up through the hole and deposited him, puffing and thrashing, onto the solid floor.

"Woo-hoo!" he panted, struggling to his feet.

"You're not allowed to fall through any floors," said Clay.

"That's definitely the last floor you'll pull me through today," he retorted. "I think we've done enough drill. It's practically inhuman in this kind of heat."

I gratefully removed my SCBA and coat, feeling the coolness of the ambient air against my sopping uniform shirt. The heat was certainly becoming oppressive. We cleaned up and made our way as quickly as possible to the truck. As I drove back to Station 8, I could see that a few clouds had gathered, but the sun still beat down mercilessly, and heat waves made the distant skyscrapers dance and shift.

My cell phone rang just after lunch, and my sister-in-law's

number showed on the display. I answered and heard Kate's voice on the other end.

"Hi, dear," she said. "We made it safe to Ann's house. She lent me her cell to call you."

"Great!" I replied. "How was the drive?"

"Oh, it was fine. I put on a CD for the girls, and they were very good."

"Glad to hear. So what are your plans for the rest of the day?"

"Well, Ann has to work at three, but I think we'll take the kids to the park before then."

"Nice. Then you and the girls will just have a quiet evening at her place?"

"I think so. We can plan something fun for when you get here. Oh, by the way, we won't be able to have our usual phone call tonight. Ann needs to take her cell to work, so this will have to count."

"Oh, well. I'll just have to sob myself to sleep tonight," I chuckled. I heard her answering laugh on the other end. "It's good to hear your voice, though," I said. "You sure we can't get you a cell of your own?"

"I'm sure. No need for our life to be ruled by these things. And we have the land line at home."

"I know, but Ann doesn't have a land line. How will you get a hold of me if there's an emergency?"

"We'll survive, for one night at any rate. Anyway, chances are slim that there'll be any kind of emergency."

"I suppose so. Well, I hope you have a nice time. I'm looking forward to tomorrow."

"Me, too. And don't forget our big getaway the day after."

"Ten years, hard to believe. We're definitely overdue

for our second honeymoon. It's good of Ann to offer to babysit."

"Oh, it'll be fun for her, too."

"Till tomorrow then. I should go. I love you. See you in the morning."

"Love you, too."

I hung up. Tomorrow was going to be a good day. We might take the children to a museum, spend time down by the water, or explore the nearby parks and walking trails. Then the next day we were planning our tenth anniversary getaway. We would be staying at an upscale hotel and going out to dinner and a movie, things we hadn't done since before we started having kids. I just had to get through this one shift, and it would be holiday time.

The afternoon grew progressively hotter, and the air hung heavy and still outdoors. In the west, fingers of haze were beginning to spread across the sky. Frank came out of his office.

"A hail warning just came up," he announced. "You guys are welcome to pull your personal vehicles into the truck bay."

As we walked out to the parking lot, I noticed that the daylight had taken on a pale and bleary quality. The haze had spread across most of the sky, and a hot wind had sprung up. A little whirlwind of loose sand traveled across the pavement for a moment, then disappeared.

"Definitely feels like it's going to storm," said Frank.

It didn't take us long to move our vehicles inside, and soon the remaining space in the truck bay was taken up with two cars, a minivan, and a pickup. Clay closed the bay doors, and as they slid downward in their tracks, they rattled a little in the wind. Through the window, I could see the trees across the street bowing and swaying. In the west, a line of cloud was growing, and there was a faint growl of thunder. I checked the time: 2:45. Hopefully Kate and the girls had left the park and were back at Ann's house by now. Suddenly, a violent blast shook the station. The windows vibrated. As quickly as it had hit, the wind died, and all was still again. The trees had stopped swaying. An eerie silence hung over everything. On the western horizon, beyond the rooftops of Lower East Side, the darkness had grown and spread, deepening to a purplish blue. For reasons I couldn't quite understand, I felt a pang of uneasiness.

Only a thunderstorm, I reminded myself.

At that moment, a dispatcher's voice came over the station radio in the watch.

"All stations, be advised. Heavy weather has hit the city. Multiple West End stations have been dispatched. South, Central, and East Stations, stand by for assignments."

"Sounds serious," said Clay.

"I'm going to listen in," I said. I hurried to the truck and switched the radio to the west end tactical channel. I was just in time to hear part of an update.

". . . And wires down at this location. We can't access the building due to fallen trees. We'll see if we can drive round and get in from the street at the rear."

This was followed almost immediately by an update from a different truck.

"Dispatch, from Engine 4. We're on scene at the corner of Highway 12 and Greenborough Parkway. We have power lines down on a row of vehicles. People are in the vehicles, and we are advising them to stay where they are. Can we get the power company to respond to this location?"

"Copy, Engine 4. From the power company, all their crews are busy at this time. We will advise them of your situation."

Other trucks were trying to make themselves heard on what was starting to sound like an overloaded channel.

"This is Safety Officer 1. My car has been hit by some kind of flying debris. Pulling over to assess the damage . . ."

"This is Car 2, I'm responding to the call at 2283 Patricia, but I will be delayed. Traffic is stalled on the freeway. Zero visibility in this rain . . ."

" . . . Power is out in the building, and the elevator is stuck between the 9th and 10th floor. Rescue 24 is trying to make contact with the people in the elevator . . . "

Then came a report that made my pulse quicken.

"Dispatch, this is Engine 21. Nearly on location at 75 North Shore Avenue. We can see that part of the roof is off. Unable to access the street due to trees and wires down. We will have to take a detour."

I remembered with a sinking feeling that North Shore Avenue was only a few blocks away from Ann's house. Quickly, I pulled out my phone. Maybe Ann had left Kate her cell. I dialed the number and pressed the phone anxiously to my ear. Ring after ring, but no answer. When it went to voicemail, I left a brief message asking Ann to have Kate call me back.

Please, Lord, keep them safe, I prayed. I knew Kate would have the good sense to get the children to shelter if she saw a storm coming, but this had come on so fast. And there were roofs off nearby . . .

The bay doors shook, and a blast of wind and rain hit the windows. I looked out and saw the trees writhing. Leaves, branches, and other debris were being hurled across the street. The cloud had increased in size as it got nearer, and it was shaped like the underside of a giant bowl. Was it my imagination, or had it taken on a greenish hue? As I watched, the wind intensified. Bushes were lying flat, and the rain was driving sideways now, thundering against the window and forcing water in under the doors. The other firefighters had gathered around, and we all stood watching the storm. The rain increased in volume, and the houses across the street all but disappeared. Raindrops striking the metal roof and sides of the station made a terrific din, and we had to shout to make ourselves heard.

"Is this a tornado?" yelled Tyler.

"I don't know," Frank yelled in answer. "Stand back from the windows, in case they break!"

We retreated into the watch. It was quieter in there, and I took out my phone again and tried calling Ann. Still no answer. A feeling of panic was beginning to well up inside me. I tried two more times. Nothing. Now I remembered her saying that she liked to leave the phone in her locker at work. If she checked it at break time, though, she might be able to tell me what was happening. I waited, staring out the window, as the others took turns calling their families. It seemed everybody else was able to get through.

"Any damage?" I heard Frank saying. "Good. Well, stay there until it dies down . . . No, we're fine here in the station . . . I'm glad you're safe . . . You, too, I'll talk to you later."

When he had hung up, I turned to him.

"Captain," I said. "My wife and kids are in the west end, where this thing first hit. If I don't hear from them soon, I'm going to have to book off and go find them."

He looked at me for a moment without saying anything.

"All right," he said finally, nodding. "Family comes first."

I knew what we were both thinking. If I booked off, it

would mean Engine 8 would be out of service until they could find a replacement for me. In the middle of a public emergency, that would be one less fire truck available to respond to people in need. It was a tough call, but if something had happened to my family . . .

"I'll give it a couple more minutes," I said. "If I don't hear from her soon, then I'll call the chief's office."

"Whatever you have to do," he said.

The rain was slackening now, and the wind seemed to have decreased in violence as well. Ragged bits of cloud were racing eastward, their fury partly spent. On the street, road signs hung crookedly. A broken tree branch lay on the sidewalk. At that moment, the printer began spitting out paper.

"Here we go!" said Clay.

I grabbed the printout and ripped it off.

"Engine 8," I read. "1405 Concession Road 10. Report of a silo collapse. Possible victims trapped."

We ran for the truck. I fired the motor with my mind racing. How could I hear from Kate now? The garage door lifted, letting in sheets of rain. I turned on the windshield wipers.

"Let's go," said Frank, buckling up his seat belt. I stepped on the gas, and we drove out into the storm.

At first, there didn't appear to be much damage as we flew through the suburbs. There were branches here and there, and shingles had been peeled off a few roofs, but otherwise the east end seemed to have escaped the worst of it. As we turned south onto Town Line Road, the houses gave way to open fields, and I noticed that a long row of power poles was leaning over the roadway. Up ahead, a wire hung low across the road. I sped under it without reducing speed. There was a loud *zing,* and the truck jolted. Instinctively, I hit the brakes.

"Keep going," said Frank.

My foot went back on the gas.

"It must've bounced off the top of the truck!" I exclaimed.

We topped a rise, and there on the right lay a ruined barn. The four walls were still standing, but the roof had been carried away. Pieces of it lay strewn all over a field of canola.

"Look at that!" exclaimed Tyler.

Worse was to come. Up ahead lay the village of Springfield. A flashing red light warned of a four-way stop. As I slowed for the intersection, it was evident that the town had been hit hard. The jumbled remains of trees, large and small, choked yards and driveways. Several had landed on houses and vehicles, smashing windshields and punching holes in roofs. Despite the rain, many people were outside, some busily moving branches, others standing helplessly, staring at the wreckage. I came to a stop.

"Turn left," instructed Frank. Fallen wires partially blocked the intersection, but I crept past them with a few inches to spare, turning onto Springfield Line. Soon we were out of the village, cornfields flying by on either side. The crops had been flattened, row upon row of corn stalks lying with their tops pointed eastward.

"Concession 10's coming up on the right," said Frank. I slowed. Just as I was beginning my turn, he stopped me.

"Wait!" he said. My foot went to the brake, and we came to a halt.

"What the—" Frank was staring down the side road. Power poles had been snapped off at the base like twigs. They lay in a neat row, one beside the other, as far as the eye could see.

"We can't get in that way," said Frank. "Stay here a minute."

He picked up his radio mic.

"Dispatch, Engine 8," he said. "We're at the corner of Springfield Line and Concession 10. We won't be able to get to that silo collapse from here. Our way is blocked by downed power lines. Is there another way around?"

We waited. No doubt dispatch was tied up with multiple communications. At last, a female voice answered.

"Engine 8. We have other trucks responding to this call from another approach. I'm going to redirect you to Sainsbury, where we have reports of several damaged houses."

Sainsbury was another village about four miles farther down Springfield Line. We drove steadily in that direction, passing another barn without a roof. As we came into the village, there was more evidence of the storm's wrath. The main street was choked by fallen wires. Trees lay everywhere, and as I slowed, I could see the naked rafters of a house that had been stripped of its shingles and roof boards.

"Looks bad down there," I said.

"Let's go that way," said Frank.

I turned, and we made our way carefully up the street, weaving around wires and branches. Most of the houses had some kind of damage. A giant spruce tree had crushed the roof of one. Another had the siding and shingles completely ripped off the windward side. The house I had seen from the main road was the worst of all. The roof was completely missing, and most of the upper floor had been carried away as well. The remains of a garage lay tumbled across the road, blocking our progress. A dismayed family stood on the driveway, gazing at the remains of their home. I stopped, and Frank rolled down his window.

"Everybody okay?" he called.

"We're okay," one of them answered. "No one got hurt."

"Do you have somewhere to go?"

"They're opening up the community center as a temporary shelter. We'll head there."

"What about your neighbors, the ones with the tree through their house? I didn't see anybody there. Did you talk to them?"

"Yes, they're fine. They said it came into the attic, but no farther. They've gone over to the community center."

"Good, well as long as you're all okay, we're going to keep checking on people."

I did a three-point turn, and we headed back to the main street. As I pulled out, we met another fire truck coming into town from the opposite direction. We stopped with our windows next to each other.

"How's it looking where you came from?" I asked.

"Lots of damage," the other driver replied. "No one hurt so far though."

"What happened with the silo?"

"No entrapment, but the farm's pretty well flattened. Owners are all right. We got sent up here to check on people."

"That's what we're doing as well."

The officer in the seat beside him leaned over and yelled across him.

"Can you guys head over to Church Street? The chief's set up a temporary command post. He'll give you assignments in person, save tying up the radio. We're going to do more wellness checks down past the community center."

"Sure thing," Frank yelled back.

I rolled up my window, and we drove on. The rain had slowed to a soft drizzle, and our tires hissed through puddles as we wove our way through the streets. At length, we came to a stop in front of an old stone building. "Our Lady of the Angels Catholic Church," the sign read. Lying in the front parking lot, as if it had been neatly plucked away and laid down by a giant hand, was the steeple. Except for the ragged edge where it had snapped off, it was completely intact, stretching its full thirty feet horizontally to the roadway.

"Holy smokes!" exclaimed Frank.

Two other fire trucks and a chief's car were in the parking lot beside it, and a group of firefighters stood around in discussion.

"Looks like our command post," said Frank. "We'll stop here for a few minutes. I'm going to rendezvous with the chief."

He got out and walked toward the parking lot. I sat in the driver's seat, looking at the damaged church. My mind was in turmoil. If the west end was as bad as this, what had become of Kate and the girls? I checked my phone. No missed calls. And I only had 3% battery life left. How could I get ahold of them?

Help me figure this out, Lord, I prayed.

Suddenly, I remembered Jane, a sprightly old lady who lived in a retirement residence not far from Ann's place. What were the chances that she would answer on a day like this? I would have one shot before my phone died. I checked my contacts feverishly. Did I have her number in my phone? Yes, there it was. I dialed and put the phone to my ear. One ring . . . two . . . three, then four. My heart began to sink. All at once there was a click, and I heard Jane's voice at the other end.

"*Hello?*" It sounded faint.

"Hi, Jane, it's Ben," I said, forcing my voice to sound calm.

"Oh, Ben, how are you?"

"I'm okay. What about you? Did you get hit with this storm?"

"Yes," she replied cheerfully. "The power's out, but we have a big generator for the fridges, and there's running water. It was a big one though, my goodness."

"I'm glad you're okay," I said. "I'm just worried about Kate. She came into town today with the girls, and they're staying not far from where you are. She doesn't have access to a phone. What is it looking like out there?"

"Well," she said, hesitating a little. "Some friends of mine from the residence went for a drive. They said the roads are clear up to the mall."

That sounded promising. The mall was only a block or two away from Ann's house.

"Much damage out that way?" I asked.

"They said they saw a few trees blown over. And also a couple roofs damaged, but not too bad. I've heard rumors that it's much worse in other parts of the city. Hard to say how bad things are really."

"It's a mess out here where I am," I said. "We'll be running all night." I stopped, my mind working. "Look, Jane," I said. "I have a big favor to ask you. Would you be able to drive over to 20 Clover Leaf Crescent? That's where she's staying. It's very close to the mall, and I think only five minutes from your place. I wouldn't normally ask, but I can't get away, and if the roads are fine . . . " I faltered. It was a lot to expect of her.

"Don't give it a thought," she said enthusiastically. "The storm's over. I'll head over there right now. What did you say the address was?"

"20 Clover Leaf."

"I'll just write that down."

There was a moment's silence.

"And Jane," I said. "My phone's about to die."

At that moment, the passenger door clicked open, and Frank climbed up into the truck.

"I'll give you my captain's cell," I said quickly. "Here, hold on . . . "

It only took Frank a second to grasp the situation. He tapped a few times on his phone, then held it out to me with his number displayed on the screen. I relayed the digits carefully to Jane, praying that my battery would not run out.

"Got it," I heard her say. "Don't worry about a thing. I'll get Kate to give you a call as soon as I find her."

"Thanks, Jane!" I said gratefully. A second later, the phone died. I breathed a big sigh.

"Found your family?" asked Frank.

"Not yet, but it sounds like their neighborhood wasn't hit too hard. I've sent someone over to check."

"All right. They're welcome to call my phone. In the meantime, we have another assignment. There are some trees that need to be cleared off the road a couple miles from here. They're blocking access to the Piper's Glen subdivision. We'll do wellness checks as we go."

We made our way through the ruined streets, stopping every so often to check on damaged homes and talk to the occupants. We were met with various reactions. One or two were in tears, others in blank-faced shock, but the majority seemed to be meeting the disaster with resolution, even pluck. The sun had broken through the clouds now, and everyone seemed to be coming outdoors. Men were up on roofs, clearing away fallen tree branches and spreading tarps over holes. On the ground, others were working with chain saws. Neighbors were helping each other dig out buried vehicles. We even drove by an impromptu lawn party, dirtied workers hailing us with raised beer bottles as we passed. Amazingly, there didn't seem to be any injuries. By the time we had left the town behind us, the wind had dropped to almost nothing, and the fields were bathed in a soft sunset glow. The last of the clouds had disappeared eastward, leaving an azure dome in their wake. Nature's fury had been spent, giving way to a flawless summer evening.

We were just coming up to the turnoff for Piper's Glen when Frank's cell phone jingled.

"Hello?" he said. "Yes . . . yes . . . he's right here." He handed me the phone. "It's your wife."

I seized the phone eagerly, holding onto the wheel with my other hand.

"Hi!" I said breathlessly. "Where are you?"

"I'm at Ann's house." Kate sounded cheerful. "Everyone's okay."

"Good to hear your voice! I was worried about you."

"It was quite the storm. We got back from the park just as it started to rain."

"That was a close call then."

"Yes it was. Of course I didn't know it at the time. It just seemed like an ordinary thunderstorm."

"How are the girls?"

"They're fine. They were enjoying watching the rain, until the fence in the backyard blew over. Then I told them to get in the basement. I think that spooked them. The power went out, but I found candles. We played games, and I read aloud to them until it was over. They're excited about their adventure now. They're already fighting over who gets to tell you about it."

"Well, sounds like they're doing fine. And I'm glad you're okay. Is Jane with you?"

"Yes, she's here. Bless her. Speaking of, I should give her her phone back. She has to get home."

"All right. Well, please thank her for me. I have lots to tell you, but I guess it will have to wait till tomorrow. I'll see you in the morning. It's going to be a busy night."

"You be careful."

"I will. I love you. Give the girls my love, too, and tell them I can't wait to hear about their adventure. I have some stories for them, too."

"I will. Love you."

I handed the phone back to Frank.

"All good?" he asked.

"All good. They stayed in the basement, and the house wasn't hit."

"That's good. Well, now that we know our families are safe, let's get to work!"

My fears dispelled, I looked forward to the night's efforts with relief, even anticipation. We carried on toward Piper's Glen. On our right, we saw what had once been a barn. It looked like it had taken a direct hit by a bomb. Nothing was left but the foundation, and the remains were strewn across several acres of field, mingled with round hay bales. Bewildered cattle roamed here and there. The silo, too, had not gone undamaged. It was snapped off at half its height, the other piece lying several hundred feet away, like a cast-off barrel. A second barn stood a few dozen yards from the foundation of the first, completely intact. The farmhouse, too, appeared untouched. A long row of pickup trucks hitched to livestock trailers was lined up along the roadside. Frank rolled down his window.

"What's going on?" he asked one of the drivers.

"Richardsons lost twenty head of cattle," the man replied. "There are about eighty more need milking tonight. We've all volunteered to take a few home till they can get a new barn built."

"You're a good neighbor," said Frank. "Any humans get hurt?"

"Not here."

We drove on. Another house with a damaged roof needed checking. The elderly couple who lived there were shaken, but unharmed. Two more stops found other occupants with plenty of damaged property, but no injuries.

"It's the strangest thing," said Frank. "All this destruction, and nobody hurt. We haven't had to do one rescue tonight!"

Not long after, we arrived at the fallen trees. Three large poplars lay across the road, blocking both lanes. I set the air brake, and we climbed down from the cab. Opening one of the compartments, I lifted out the chain saws, fired one up, and walked toward the trees. Tyler grabbed the other one, and together we attacked the tangle of greenery, cutting away limb after limb until we had reached the bare trunk. Frank and Clay worked behind us, pulling the severed branches out of the way and stacking them in the ditch. We bucked up the tree trunks into firewood-sized lengths, which were then rolled off to the side. When the last piece had been cleared away, I shut off my saw.

"That road's open now," said Frank with satisfaction. "They'll be able to get an ambulance into Piper's Glen if needed."

Soon we were driving back toward the command post for our next assignment. Before we got there, however, a voice came over the radio.

"Engine 8 from Dispatch."

"Go ahead," said Frank.

"I need you to return to your district. We anticipate multiple alarm calls as the power comes back on. We will send them to you in order of priority."

"Copy, Dispatch. We'll return to station."

"Going to be a busy night," said Clay.

"Yes," said Frank. "It's not over yet, that's for sure."

Dusk was falling as we turned west onto Springfield Line, heading back in the direction of the city.

It was completely dark when we got to the station. We just had time to gulp down a few leftovers before the first of the calls came in. Power had been restored to a number of city blocks, and fire alarms were going off everywhere. There were no fires, but something about the storm had triggered several detectors, and an already-overwhelmed dispatch center was being flooded with more calls. By midnight, my eyes were blurring as I drove, and conversation lagged in the truck. Luckily, the calls petered out around one in the morning, and I was able to snatch a couple hours of sleep in the watch. It proved to be only a lull, however, and before long we were back on the road, responding to more alarms. As the sun came up, we pulled into the station for the last time, eager to pass the torch to the oncoming crew. My voice slurred a little as I gave my morning debrief to the B Shift driver. His eyes widened a little as I described yesterday's events.

"We'll be doing lots of cleanup today," he observed.

I swallowed a hasty cup of coffee, enough to keep me awake for the drive, and bolted for the door. At last the long shift was over. A few minutes later, I was merging onto the freeway, heading westbound. As I drove, my eyes took in the altered landscape. The high rises were unchanged, but here and there, in the parks and green spaces where trees had once stood, the skyline looked strangely bare. Line crews were hard at work, bucket trucks extended, trying to restore power to the most devastated areas. It would be days, I thought, before they could work their way through the whole city. After twenty minutes of steady driving, I took the exit for the west end suburbs. At the end of the ramp I slowed and looked out the window with amazement. On my right lay the remains of a high-voltage tower, its giant steel frame twisted, its sturdy cross-beams crumpled. Under a tangle of fallen wires were six abandoned vehicles. If this was the scale of damage, I realized, it might be weeks, not days, before we had power again.

Minutes later, I was knocking on the door of 20 Clover Leaf Crescent. There was a noise of stampeding feet, and all at once the door burst open, and three little girls were mobbing me with rapturous cries of "Daddy!"

Kate was close behind them, and she fought through the throng to give me a welcoming hug and kiss. I held her close for a minute, a feeling of gratitude welling up inside me.

Thank you, Lord, I prayed. *Thank you for keeping them all safe.*

Ann stood in the background, smiling at the noisy reunion.

"You're here at last," said Kate, with a tired but happy smile of her own. "And you're just in time for breakfast."

There was no power, of course, but we sat down to a hearty breakfast of peanut butter sandwiches and apples. The children chattered enthusiastically, vying with each other to tell all about their adventure.

"Were you scared?" I asked.

"Yes!" they said, nodding together emphatically.

"But only at first," Beth added. "Mommy was there, and we had *lots* of fun. Oh, Daddy! We get to tell Grandma that we were in a *real* tornado!"

When the meal was over, I turned to Kate with a wan smile.

"Tired?" she asked.

"I was thinking of our getaway," I said. "I'm afraid not many hotels are open, and the few that are will be putting up people who have lost their homes."

She sighed. "No dinner at the Hilton. But I can't feel sorry for myself. Not with so many people in need."

"Ah, who needs steak dinner?" I said. "We have peanut butter sandwiches!" I looked at the happy children. "And the best company in the world," I added.

17

TWO PATHS

"Engine 8, from Dispatch."

"Go ahead," said Frank.

"Engine 8, be advised, this is for a 38-year-old male collapsed on the driveway, vital signs absent. Covid screening is positive."

"Covid protocols, boys," said Frank over his shoulder. Tyler and I exchanged looks, then began donning our extra layers of protection. It was never easy working a medical call in full firefighting gear. I slipped the straps of my SCBA over my shoulder and donned my mask. Next, I struggled into a medical gown, pulling the SCBA free of its bracket so that I could turn my back to Tyler. No words were necessary, we both knew the drill. He grabbed the strings of my gown and tied them together at the back, over the top of the air tank. Then he turned his back to me, and I did the same for him. We were just pulling on our double layers of med gloves when the truck arrived on the scene.

Grabbing the first-aid kit, I hurried toward the human figure lying on the driveway, Tyler following closely with the defib and oxygen. A heavy-set man in sweat pants and hoodie lay face-up beside a car. His eyes were open, pupils fixed and dilated, staring lifelessly at the sky. I did a quick pulse check and shook my head.

"Definitely VSA," I said, my voice muffled through my mask.

"What?" said Tyler.

"No pulse!" I yelled. *Stupid masks,* I thought.

"Copy!" Tyler replied, putting down the equipment.

"Start CPR!" I told him. "I'll set up the defib!" He began compressions, and I worked around him, snipping open the patient's shirt and placing the defib pads on his bare chest.

"Analyzing," the machine announced in its electronic voice, and Tyler paused, keeping his hands free of the patient.

"No shock advised," came the prompt. He resumed compressions. Frank came up behind us, giving a radio update to dispatch. I took control of the head, clamping the bag-valve mask firmly over the patient's nose and mouth and giving the bag a squeeze every five seconds to force oxygen into his lungs. When Tyler had finished three minutes of CPR, we switched. Kneeling down beside the chest, I felt for the sternum and began pushing. My SCBA tank clunked awkwardly against my back as I worked, and I fought to pull breaths of compressed air out of my facepiece. Tyler gripped the oxygen mask firmly, his own breath sounding labored after his three minutes of exertion. Another three minutes went by. A second machine analysis indicated no shock advised, and we switched positions. A siren was wailing somewhere in the distance. The ambulance was on its way. Behind us, I could hear a woman's voice addressing Frank.

"I found him like this when I came out to start my car. I have no idea how long he was lying there."

"You live next door?" Frank was asking.

"Yes," she replied. "My husband and I hadn't seen him in a couple of days. He was sick."

"Covid?"

"I don't know for sure. He was worried that he might have had an exposure last week. I told 911 that he might be contagious."

"Better safe than sorry. Were you in contact with him yourself?"

"We kept our distance. Here, I'll make sure I stay back six feet from you as well . . . Also, you should probably know he had a bad heart."

"That's good to know, thank you."

The siren increased in volume, then shut off. An ambulance turned the corner and cruised to a stop in front of the driveway. Two paramedics climbed out, dressed in medical gowns, surgical masks, and clear plastic face shields.

"What have we got?" one of them asked.

"Thirty-eight year old male, history of heart problems, confirmed VSA," replied Frank. "We've done two cycles, no shock advised. We're counting him as Covid positive."

"Let's get him in the ambulance. Can I borrow one of your guys for compressions?"

"Clay, you go," Frank ordered. "You're fresh."

Clay come forward and took over CPR, while Tyler and I assisted the paramedics in loading the patient. I had a last sight of Clay pushing hard on the patient's chest as one of the paramedics hooked up tubes and wires. The driver of the ambulance closed the back door, then remarked wryly:

"That'll count as another Covid statistic."

In a minute, they were gone, and Tyler and I stood on the street drenched in sweat.

"We'll do our decon protocol by the truck," said Frank. Together we walked over to Engine 8, and the captain opened a side compartment and got out a spray bottle and two yellow bags marked with the biohazard symbol. We stood still while he sprayed us from head to toe, the antiviral agent leaving a pungent-smelling mist hanging in the air. After that, we waited for ten minutes to allow the chemical to kill off any germs. Frank checked his watch at intervals.

"That's enough," he said at last. He held one of the yellow bags in front of me, and I pulled off the medical gown, stuffing it in as far as I could. The gloves followed. Tyler took off his layers next, while I removed my SCBA and placed it carefully into a clear plastic bag. It would be scrubbed with soap and water back at the station. After squirting a generous amount of hand sanitizer onto our open palms, Frank gave us the nod, and we climbed aboard the truck. I took over the driver's seat, and we drove to the hospital to pick up Clay. He was waiting for us outside the emergency ward. Tyler and I sat in the truck while Frank performed the same decon rituals on Clay. Finally, his ten minutes had elapsed, and we were on our way back to Station 8.

"How was the patient?" asked Frank.

"No go," said Clay blandly. "They pronounced him dead not long after we got there."

When we got back to the station, Clay backed the truck into the bay, and I climbed down and peeled off my bunker pants. Tyler scrubbed the SCBAs on the bay floor while I sprayed down the defib, oxygen case, and first-aid kit with disinfectant. I cleaned the inside of the cab thoroughly with sanitary wipes, sprayed the bottom of my boots, then headed to the dorm to shower and change my uniform. The old clothes would go in a plastic bag to be washed at home, separate from household laundry. When I emerged, Tyler was sanitizing the kitchen, a daily routine we had followed since the start of the pandemic. He was heating a large pot of water, into which he dumped half a cup of bleach. All the silverware and utensils followed, where they would boil for half an hour.

"Ben, can you do the morning wipe down?" Frank asked, pausing at the sink where he had been lathering his hands.

"Sure," I replied. I went to the janitor's closet and got

out more sanitary wipes. I worked from room to room, disinfecting counter tops, light switches, phones, computer keyboards, door handles, and anywhere else people might have touched. That done, I returned to the kitchen to help Tyler put the utensils away.

Clay was busy with the groceries. Our meals had to be picked up on our days off now, since Covid regulations forbade crews from entering public spaces unnecessarily. It had been Clay's turn to shop, and now he was preparing his latest master plan: homemade shawarma. I helped chop the vegetables, while Tyler spiced the chicken and took it out to the barbecue. In less than an hour we were sitting down to an unusual but delicious lunch. The white meat was cooked to perfection and cut into thin strips. Platefuls of pickles, sugared onions, shredded lettuce, and diced tomatoes filled the table, accompanied by three kinds of sauces. We sat down at our usual places around the table. Technically, we weren't supposed to eat in the same space, but Frank took a common-sense view of things and allowed us to have meals together as normal. I scooped each ingredient carefully onto a pita, rolling it up into an oversize wrap. Frank took an enormous bite of his and exclaimed:

"Dang, that's good!"

"It's all right," said Clay.

"What's going on in the world?" asked Tyler, aiming the remote at the television. A newscast flashed across the screen.

"Same news," said Frank with his mouth full. "Guess what? There's a pandemic happening."

"I wish they'd talk about something else," said Tyler.

"Here's some good news, though," said Clay, who had been following the dialogue between the talking heads on-screen. "Looks like the vaccine should be here soon."

"It's about time!" said Frank loudly. "I wanna get out and do things again."

"Yeah," added Tyler. "I want to go on my Caribbean trip. I had to cancel it last year."

"How soon do you think they'll allow travel?" asked Clay.

"Hard to say," said Frank. "They'll probably want a certain percentage of the population vaccinated before they open things up."

"You'll definitely need it to fly," put in Tyler. "They're already talking about vaccine passports."

"Makes sense to me," said Frank. "We have to stop this thing somehow."

I didn't say anything, but I felt a certain sense of uneasiness. This talk of vaccine passports had an Orwellian ring to it.

Late in the afternoon, the phone rang. I recognized the voice of Chief MacLean on the other end.

"Can you inform your captain that there are some changes coming?" he began curtly.

"Sure, Chief," I said.

"Due to Covid, we're going to have to move some personnel around," he said. "Station 16 has ten firefighters, which means they can't do social distancing in their station very well. We're moving Rescue 16 to Station 8 next shift."

"Oh, wow!" I said. "Okay, I'll let the captain know."

"You'll have seven people there," the chief went on. "Four on the engine, three on the rescue. You may have to rearrange the station a little bit."

"Copy that."

"Thanks."

He hung up. I hurried to Frank's office and knocked on the door. When I told him the news, he seemed surprised but not displeased.

"So we're going to have company," he said warmly. "Well, the more the merrier, as far as I'm concerned. You guys can set up a couple extra beds in the dorm, and the spare bedroom can become the lieutenant's office."

We spent the rest of the afternoon preparing for our guests. Two spare cots were set up beside the two existing ones. With the driver's berth in the watch and one bed in each of the officers' rooms, there would be beds enough for everyone. A table was borrowed from the training room to extend our dining space. Various stored items were cleared off the bay floor to make room beside Engine 8 for Rescue 16. When all was ready, Frank looked over the new arrangements approvingly.

"The invasion begins next shift," he said. "I just checked the roster. We'll have Bryce Halton here as lieutenant. The two firefighters will be Brian Dervish and Chris Hallahan."

"Every shift?" I asked. I knew Brian, a bit of a loudmouth, but not impossible to get along with. Chris I liked. He had been a combat medic in Afghanistan and hid a razor-sharp skill set behind a kindly, soft-spoken demeanor. I felt my heart sink at the mention of Bryce.

"I'm not sure how it'll work," Frank replied. "I think the captain at 16 may rotate people through regularly. We'll see."

"How strict are the Station 16 boys about masks?" asked Tyler. "Will they expect us to wear them all day?"

According to the new regulations, we were required to wear surgical masks on duty at all times. Frank scoffed at this measure, and only told us to put them on if the District Chief or some other unwelcome visitor happened by.

"It's my station," he now observed. "I'm not going to make anyone wear a mask if I can help it. I can't speak for the rest of them, but I know Bryce doesn't think much of all the Covid hype."

I silently filed this last bit of information away. I couldn't help feeling nervous about working with Bryce again, but now it sounded like we at least had a small thing in common.

I arrived for work the next shift with a tingling of apprehension. I could already see through the bay door windows that there were two trucks parked in the station. I entered the man door and began the daily Covid screening procedures. On a table just inside the entrance was a stack of papers, a box of surgical masks, hand sanitizer, and an infrared thermometer. Picking up the thermometer, I pointed it at my forehead, holding it steady until I heard it beep, then recorded my temperature in the logbook. The numbers were in the acceptable range. Next I read over all the boxes: fever, sore throat, new or worsening cough, travel outside of the country in the last two weeks, contact with any symptomatic persons in the last ten days. I checked "no" to all the above and sanitized my hands. Then, ignoring the masks, I walked into the station to start my day.

I was just setting up my gear on the rig when Bryce came around the corner of Rescue 16. He greeted me with a polite nod.

"Hey, Ben, how's it going?" he said, in a not unfriendly tone.

"Not too bad," I replied. "Congrats on moving up to acting lieutenant."

He gave a small chuckle. "Thanks," he said. "Took me long enough, but here I am."

He walked into the living quarters. After finishing my morning checks, I followed and joined him in the kitchen.

The rest of my crew was there also, with the two firefighters from Rescue 16. The kitchen seemed smaller with so many bodies, and the space rang with loud conversation. Tyler was busy at the stove, whipping up a double batch of scrambled eggs. I shook hands with Brian and Chris. Neither was wearing a mask.

"Welcome to the Great Eight," I said. "You guys going to be here permanently?"

"More like the Crazy Eight," rejoined Brian with a grin. "Naw, it's only going to be for a month, then we'll go back to 16, and they'll send someone else down."

"I probably won't be here the whole time, either," added Bryce. "There are a few other senior firefighters who will need acting experience on the rescue."

"How long is the rescue going to be here?" asked Frank.

"Hard to say," he answered. "Maybe for the rest of our lives. Apparently, this 'pandemic' is never going to end, at least as long as the media have their say. It's too good an opportunity for government control."

He gave this last opinion with a bark of a laugh, but there was an uneasiness in his eyes. Clearly, he was not accustomed to a receptive audience. None of the others replied to this, and I thought I saw Brian roll his eyes.

"There's a lot of fearmongering," I ventured. Bryce turned and looked at me for a moment.

"Fearmongering, exactly," he said, without shifting his gaze. "They want us to believe we're all going to die, unless we follow the rules."

"Well, I don't want to speak against legitimate safeguards," I hastened to explain. "But some of these measures do seem like overkill. I'm thinking of the little kids who are forced to wear masks all day in school."

"Oh, I know, it's crazy," he agreed, warming. "And so

unhealthy for them. You know, if you look at history, every totalitarian regime has done the same thing. Start by making everyone afraid, and then they'll be willing to hand over their freedoms. You can get people to do anything you want after that. I'm telling you, this is the beginning of the end of our democracy."

He relapsed into silence. I wasn't sure if I agreed with him, but there was a keenness to his insight, and it had something in common with my own suspicions. Frank quickly changed the subject, and for the remainder of breakfast Bryce said little, listening while the others talked about sports and the weather.

After breakfast and truck checks, I decided to send an email to Ted English. "Dear Mr. President," I wrote. "There has been a lot of talk in the media and in the workplace about Covid vaccines. What is this going to mean for firefighters? Will we be expected to get vaccinated? I am concerned about the serious side effects that have been reported in various case studies. I am anxious to know what position the union will take on this issue. Can employees be forced to take a medical product that is experimental?" I hit "send". Union officials could take quite a while to respond to inquiries, so there was nothing to do now but wait. I had a spare hour, so I decided to see if the Catholic Church had anything to say about it. A quick internet search brought up a guidance note by the Congregation for the Doctrine of the Faith, one of the Vatican's most eminent committees. I read with interest as it laid out the criteria for Catholics to follow when choosing a vaccine.

> The question of the use of vaccines, in general, is often at the center of controversy in the forum of public opinion. In recent months, this Congregation has received several requests for guidance regarding the use of vaccines

> against the SARS-CoV-2 virus that causes Covid-19, which, in the course of research and production, employed cell lines drawn from tissue obtained from two abortions that occurred in the last century. At the same time, diverse and sometimes conflicting pronouncements in the mass media by bishops, Catholic associations, and experts have raised questions about the morality of the use of these vaccines.

The document went on to explain that, while Catholics have a firm duty to refrain from cooperating in an evil action such as abortion, there also exists what is called "remote material cooperation" in an evil that has already occurred. This may or may not be wrong, depending on the circumstances. In most cases, Christians should resist receiving a product such as an abortion-derived vaccine, because they would otherwise help create a demand for that product. However, in cases where there was a legitimate and grave need, and no licitly made alternative was available, it could be acceptable to make use of it.

In the case of Covid vaccines, then, the Church allowed that it would be acceptable to receive an abortion-tainted shot, given the seriousness of the pandemic. However, doing so would come with serious responsibilities. First, Catholics must disapprove of the use of abortion-derived products and strive to make ethically produced ones available. The document then made an appeal to governments to oversee the development of pharmaceuticals that do not create a crisis of conscience for people of faith. Near the conclusion, it stated that people should never be forced to take a vaccine, and that the choice to receive a medical product lay with the individual, not the state.

I read the whole thing again, carefully. It was hard work wrapping my head around the nuances, but a basic impression was emerging. So, the Church didn't want me to support the abortion industry, but if things got bad enough, she would allow me to get an abortion-derived vaccine, as long as no better ones were available. Most importantly, it would be wrong for my employers to force me to it.

Now it remained to be seen which vaccine brands were acceptable. I did another lengthy search, pulling up as much information as I could about the pharmaceutical companies currently working on a Covid vaccine. As it turned out, of the four brands soon to be available to the public, two of them made use of fetal cell lines at the testing phase, and two actually used fetal cell lines in production.

I logged off, my head spinning. It was a lot to process. I sighed. I could only hope that management would not make this compulsory.

We pulled up on the scene and saw that it was a head-on collision. Engine 15 was there ahead of us, and the crew was already stabilizing the vehicles and tending to the one patient still trapped. Tyler and I grabbed the extrication tools and hurried to the mangled car, just as Rescue 16 pulled up behind us. Tyler wielded the spreaders, working away at the driver's door, while Bryce came up behind him to give direction. I waited nearby with the cutters, ready to shear the door off its hinges as soon as the latch side was free. It was stubborn, and I could see the frustration on Tyler's face as he fought with it. Thirty seconds went by, then a minute. Bryce gave orders patiently. At last it popped off of the latch. Losing no time, I crouched by the hinges, freeing the door with two quick cuts. In another three minutes, an elderly woman was being extricated from the car, in serious but stable condition. We helped the paramedics wheel her to the ambulance, and moments later they were bearing her away. Bryce gave me a brief pat on the arm.

"Good job," was all he said.

Later that day, I checked my email, and found a reply from Ted English in the inbox. It was brief and to the point.

> Dear Ben,
>
> Thank you for your inquiry. The vaccine is not mandatory at this time. We anticipate that it will be available soon, and all members are encouraged to get vaccinated. The union has advocated for firefighters to be included in the first wave of vaccinations, meaning that our members will be among the first to be eligible for the shot. At this time management has not indicated whether they will create a mandatory vaccination policy, but they may do so in future.
>
> Sincerely,
> Ted

This was good news, of a sort. If I was forced to make a decision, at least it would be farther down the road.

The station telephone rang a week later. Frank came into the kitchen where we were all assembled.

"There's a vaccine clinic happening today at Central Hospital," he announced. "The chief wants the names of everyone who's going, and we're supposed to arrange transportation. If there's enough interest, both rigs will go. If only a few of you want to go, then we'll just take the engine, and leave the rescue to cover the district."

There was a short silence.

"Is it mandatory?" asked Chris.

"No," said Frank. "Highly encouraged, but not mandatory. I'll be going for sure. Anyone else, just let me know."

"I'm going," said Brian emphatically. "Not sure why anyone wouldn't," he added.

"Me, too," Tyler said. "I got my trip to the Bahamas coming up."

"Count me in as well," said Clay.

There was another pause. Bryce had a grim expression on his face, and Chris looked uncertain.

"Anyone else?" prompted Frank.

"I'll pass," I said.

"That's fine," said Frank. He look surprised, even disappointed. "Bryce?" he asked.

"Hell, no," the lieutenant answered. Frank made no comment, but his face reddened a little. He turned to Chris. "What about you, Mr. Hallahan?"

"I'm on the fence," Chris began hesitantly. "I think I'd like a little more time to decide, to be honest."

"Okay, buddy, no problem," said Frank. "These clinics aren't going away. We have four people, and there are four seats in the engine, so that works out well enough. Ben, you can hop on the rescue for the afternoon."

"Sure thing, boss."

When they had gone, I decided to probe Bryce's thoughts.

"So, no vax for you, either, huh?" I began. He shook his head.

"Since when do they push a medical product on people that's still a hundred percent experimental?" he said, a note of intensity in his voice. "Sure, if it's the cure for Covid and it won't give me any side effects, sign me up. But we don't know that yet. And all this hype! They're putting people under so much pressure to go out and get this thing, and we don't know what it's going to do to us."

"There is that," I said. "I think it's the coercion factor that bothers me the most."

"You watch, it'll be mandatory before long," he said darkly. "For everyone, not just nurses and doctors. They were forcing medical experiments on people in Nazi Germany. This is just the beginning."

"It could be," I said cautiously. "I think totalitarianism would look very different here than in Germany. But you're right that there are changes happening. We're not the same society we were even thirty years ago."

Chris had been a fly on the wall during this exchange, and now he spoke up.

"I don't know if I agree with you guys about the way things are going," he ventured. "I just want more information before I get vaccinated. I'd like to see some studies, long-term effects, that kind of thing. In the military, there were a bunch of guys who had to get a shot before going overseas. I forget what it was exactly, but they complained because there was some kind of ingredient that caused neurological problems. Well, they were threatened with court-martial if they didn't comply, and now twenty years later some of them are really sick. There's a big lawsuit over it. That's why I'm leery until I know more."

"If you want more information," said Bryce, an eager, almost fanatical light coming into his eyes, "you should check out all the data on vaccine injuries. They're in the thousands for deaths! Tens of thousands for other problems!"

Chris looked dubious. "You'd think that would be on the news," he said.

"You think the news wants to cover this? They're eating out of the government's hand! And the government wants these vaccines!"

"I dunno," said Chris skeptically. "It sounds like a conspiracy theory to me."

"Well, there are public websites that publish the facts,"

Bryce replied. "You can read about it for yourself. But that's only the tip of the iceberg. There's every reason to believe the medical world is obscuring data to stop people from knowing how bad these vaccines are. Case in point, I was talking to my doctor the other day. He told me that whatever I do, don't let my kids get the shot. He said it's absolutely terrible for them. He also told me that if his colleagues got wind of his opinion, he'd probably lose his license."

"Come on," said Chris. "Are we really at that point?"

"If we aren't, we're getting there. We should all be very concerned for our freedoms. The problem is, most people are totally complacent."

"I still don't know," said Chris, shaking his head. "You may be right about all that, but at the end of the day, I have my job to think about. If they make it mandatory . . ." He didn't finish the thought. It was the big question that was on all three of our minds. If they made it mandatory, what would we do?

The news came about six months later. It was the afternoon of the Riverview Hotel fire, and we were all exhausted after a long morning of firefighting. Cleanup was done, the post-fire debrief was over, and I had showered and changed into a fresh uniform before sitting down at the station computer. A mass email was in my inbox.

> Attention: All City Staff,
>
> In response to the growing number of Covid cases in our community, the city will be implementing a mandatory vaccination policy. Every employee must submit proof of vaccination against Covid-19 in order to retain employment. The city will consider requests for exemption based on health or disability. However, employees eligible for

> exemption will not be able to continue in a paid position. The policy will allow for members who are unable to get vaccinated for legitimate reasons to take an unpaid leave of absence. (Note: Employees placed on leave may not be able to retain their position with the city indefinitely.) Any employee not in compliance with this policy will be subject to disciplinary action, including termination of employment.

It was signed by the city manager.

At home, I sat down at the kitchen table with Kate and told her the news.

"It's here," I said.

"I think we knew it was coming," she replied quietly.

"What am I going to do?" I asked. "I don't want to get this shot. Not yet at any rate, not until we know more, or at least until there is an ethical brand available. But we're out of time."

"We need to pray," she said.

"Yes," I agreed. "Pray, and get advice. The problem is, there are so many voices. The Church hasn't exactly said it's wrong. And I know many good people who have gotten vaccinated, including people I look up to, priests even. I guess it comes down to each person's conscience. And that's the hard part. I wish it were more clear-cut."

"You'll come to the right decision," she said encouragingly. "You always do."

She was supportive, as always, but I felt the burden resting heavily on my shoulders. I didn't relish the idea of getting an abortion-tainted shot, but I had a family to feed. I knew I was at a crossroads. Then, dimly at first, but gaining clarity every minute, a half-forgotten memory emerged. It was fall. My older brother and I were on a boyhood excursion through the woods, and the leaves hung gold and crimson from the maples and birches. We were following an old deer path, and ahead of us lay a fork in the trail. My brother, who was leading the way, paused.

"It's the poem," he said.

"What poem?" I asked.

"You know, the Robert Frost poem," he replied insistently. "The one I had to memorize for school."

"I don't know that one," I said, impatient to move on. But my brother stood still, gazing ahead of him at the divide. All at once he began to recite:

> Two roads diverged in a yellow wood,
> And sorry I could not travel both
> And be one traveler, long I stood
> And looked down one as far as I could
> To where it bent in the undergrowth;
> Then took the other, as just as fair,
> And having perhaps the better claim,
> Because it was grassy and wanted wear

I couldn't remember the rest, but the image of the fork in the trail had stayed with me for many years, until at last time took its toll and the memory faded. Now, the image was revivified, given a sharp new meaning by my present crisis. Which should I travel, the well-worn path or the lonely and more difficult one?

I took the matter to prayer again and again over the next several days. In moments of silence, I could almost grasp the answer. "No, this is not for you," seemed to be the recurring theme, but the reasons for this answer kept eluding me. There were many voices competing for my attention. The problem was, I had friends and family on both sides of the debate.

"You need to protect your loved ones," seemed to be the

dominant rationale for those in favor of the vaccine. "You'd never forgive yourself if you gave the virus to someone and they died."

"What about your career?" was another question posed by concerned relatives. "How will you pay the bills if they take your job away?"

As compelling as these reasons seemed at times, something continued to hold me back. On the other side, there was a whole spectrum of theories, ranging from paranoid conspiracy theory to pseudo-mystical conjectures about the end times and anti-Christ. It was true that my concerns shared a nexus with the tamer end of the spectrum when it came to these ideas; still, I wasn't sure that I wanted to be lumped in with the "lunatic fringe". At the same time, there was a growing concern on Kate's mind based on what she was hearing from our circle of friends and acquaintances.

"Your mom phoned this morning," she told me one afternoon.

"Oh, yes?" I said.

"There's been another death."

"Who?" I asked sharply.

"The brother of one of their friends from church. Apparently, he was a healthy, thirty-something-year-old dad. Six kids. He had to get the vax for work. Within a week he was dead from a heart attack."

"How many is that now?" I asked, pursing my lips. "I mean, friends of people we actually know? I'm not counting the stories on social media."

"At least a dozen," she said. Her face had a pained expression. "Ben, I don't think we can ignore this anymore. If you get vaccinated, I don't know if I can live with the worry that I might lose you. No job is worth that. I mean, how close to home does it have to hit before we're willing to say no?"

"These things could be coincidence," I said slowly. "There is really no way of proving that it's because of the vax."

"This many coincidences, in this short a span of time?"

"I just don't know what to think," I said. "All my friends who have taken it are doing just fine. These are second-hand, third-hand stories . . . Do we base a big life decision on them?"

"Can we afford not to?" she asked.

I didn't know what to make of all the conflicting information, but two points at least stood out clearly to me: Coercion of workers by management was undemocratic, and the vaccine's link to the abortion industry troubled me. If I was to take a stand, I decided, it would be important to get the facts on abortion-derived cell lines. Based on what the pro-life agencies were saying, these products had been used to develop every brand of Covid vaccine on the market, morally tainting each batch and causing an ethical dilemma for Christians considering whether to get vaccinated. But perhaps, I reasoned, the cell lines didn't actually stem from an abortion. Maybe certain pro-lifers only worried that they were. Maybe they were from a child lost to miscarriage, whose body had been donated to science. I would have to do some extensive digging, I thought, to get to the truth of the matter.

As it turned out, I didn't have to look far. An article citing research from the 1970s revealed the origins of the cell lines that are derived from human tissue. What I read nearly made my stomach turn. According to statements made by a physician who had worked on developing one of the earliest lines, a perfectly healthy baby girl had been aborted in the Netherlands. Her kidneys were extracted and cut into pieces, which were then used in the cloning and development of a virtually inexhaustible bank of human DNA known as the HEK293 cell line. The pro-life movement had named the baby girl Johanna, meaning "God is gracious", posthumously giving her an identity and a dignity. All that her so-called doctors had given her was a numerical label and a grave in the trash. I shook my head in disgust. So this was what they were forcing us to participate in. And yet, even knowing this sordid history, I still had my family to think of. If I took a stand on this one, we could end up on the street.

If anyone could help me, I thought, it was John Paul II. I knew he had written extensively on bioethics and that among his writings were papal encyclicals, which had moral authority. I spent the next few hours poring over his various works, drinking in the wisdom of his insights on the origins of life and the dignity of the human person. There was a wealth of information to unpack, and my head was beginning to ache. Still, I felt no closer to an answer. Then all at once a passage leapt out at me. It was a simple paragraph, not particularly eloquent, but it spoke right to the heart of my dilemma: "The passing of unjust laws often raises difficult problems of conscience for morally upright people with regard to the issue of cooperation, since they have a right to demand not to be forced to take part in morally evil actions. Sometimes the choices that have to be made are difficult; they may require the sacrifice of prestigious professional positions or the relinquishing of reasonable hopes of career advancement." I stared at the page, nodding and nodding. I knew what I had to do.

"Kate," I said, as she came into the room, curious as to what had kept me ensconced for so long. "Listen to this!"

I read her the passage.

"I've made up my mind," I said. "I'm going to fight this."

18

THE ROAD NOT TAKEN

"How are your ribs?" asked Frank, putting down his coffee cup.

"They're fine," I answered quickly, concentrating on my morning eggs.

"Ribs?" asked Tyler from the stove.

"Oh, you weren't here last shift," said Frank, a grin spreading over his face. "Have I got a story for you!"

"Here we go," I groaned.

"I wanna hear a good story!" said Brian, looking up from his cell phone. Chris paused his breakfast and grinned at me good-naturedly.

"So, it's three in the morning," Frank began. "I get a knock on my door. There's Ben, looking all worried. 'Captain,' he says. 'I think I need to book off. I'm having chest pains.'"

"Wait, wait, wait!" I broke in. "Back up a little. Let me tell what happened first!"

"Fine," he laughed. "You tell the first part."

"Okay, so," I said hastily. "What the captain left out was that we had gotten a call earlier. Just the engine. Dispatch said there was someone unconscious, locked into a washroom at the coffee shop on Glenvale. So we show up, there's like three cop cars there and an ambulance. We go in with our forcible entry tools. Cops are all standing around, paramedics with their stretcher, waiting for us to break the door down. The owner is all riled up, keeps telling us the door's locked and he's sure someone's in there. So Clay and I go to work with the Halligan. The door won't budge. I'm telling you, it was the toughest door I've ever forced."

"It was a solid steel door," put in Clay. "We must've been going at it for at least five minutes."

"Yeah," I agreed. "Finally, I got mad and picked up the sledge hammer. I started reaming on it. I don't know how many times I hit it, but it finally gave out."

"You should've seen the mess," said Clay. "Smashed tile everywhere. There was a whole piece of wall gone. And the door was a write-off. He must've done about a thousand dollars' worth of damage."

"So I look inside," I went on. "Nobody there."

Tyler broke into an amused smile. "No one?"

"No one. We destroyed the poor guy's bathroom for nothing. I have no idea how that door was locked with no one inside it. There was no mechanism on our side."

"Weird!" said Brian.

"So, we get back to the station around eight at night," said Frank, taking over the narrative. "Everything's normal. Then I get woken up by Ben, thinking he's going to have a heart attack."

"It really hurt," I said defensively.

"Well, you looked worried enough. Tell the guys what the doctor said to you at the hospital."

"Fine, if I have to. Well, first of all, they did a cardiogram," I said. "I told them I'd had chest pains a couple of years ago, but nothing like this. The doctor came back and said, 'There's absolutely nothing wrong we can find with your heart. Were you in an accident or something?' I told him no. 'Some kind of impact or sports injury?' he said. Then it clicked." I tapped my head with a wry grin. "I told him I'd been whacking a steel door with a sledge hammer. He said, 'Yes, that would do it. You've damaged the cartilage in your rib cage. Unfortunately, there's nothing we can do for that. You'll just have to go home and rest for a few days.'"

Everyone was laughing now.

"I wish I'd been there to see it," Tyler grinned. "Big, lanky farm boy raging the door down. Still, I'm glad it wasn't a heart attack."

"Me, too," I said.

"That's why I didn't see you the next morning," said Chris. "Are you okay?"

"I'm fine," I replied. "It doesn't even hurt now."

I'm going to miss this, I thought. I had come to love this part of the job: the storytelling, the banter, the brotherhood surrounding me each day. I wondered what the guys would think if they knew that this might be one of my last shifts. The deadline for vaccination was looming, and if I couldn't get a favorable hearing from the chief, there was every reason to believe that my days at Station 8 were numbered. I fought down a feeling of apprehension. My interview with him was scheduled for tomorrow afternoon.

Partway through the morning, I met Chris coming out of the watch.

"I just took a phone call from Chief MacLean," he said. "The rescue's going back to Station 16 tomorrow."

"Really! Did he say why?"

"Case numbers are down. They're not so worried about the virus any more."

"They're not worried," I said. "But they'll kick out anyone who's not vaccinated!"

"I know," he said. "It doesn't make sense. I went and got my first shot," he continued, confidingly. "I had to think of my kids. Things are a bit tight right now. Can't afford to lose my job."

"It's a tough spot to be in," I said sympathetically. Inwardly, I wondered if I should be doing the same thing. Was I crazy to be going against the flow?

Next morning, I stood in the lobby of fire headquarters, waiting for my meeting with the chief. There were butterflies in my stomach, and I had to practice deep breathing to keep my rising agitation at bay. Rank and file firefighters rarely frequented the chief's office, and then only when they were in deep trouble. I remembered my conversation with John Forbes over two years ago. There had been genuine sympathy, as well as professionalism, in the way he had handled my case. Since then, he had gone on to become Chief of Department, and now he held the reins of power for the entire organization.

I heard footsteps. Someone was bounding down the stairs. The door opened, and Chief Forbes sprang lightly into the room. He had five gold bars on his shoulders now, but he looked otherwise unchanged. There was a warm smile of welcome on his face.

"Ben!" he said enthusiastically. "Come on in."

I stepped forward and shook his hand.

"Thank you, Chief," I said. "It's good to see you again."

"You, too," he replied. "How about we head up to my office?"

I followed him up the stairs.

"How is your day going?" he asked.

"Pretty good," I answered. "I just got off shift this morning. How about yours?"

"Well, you know, a chief's life is always busy."

He ushered me into a small room, bare except for a round table and two chairs. He shut the door.

"Have a seat," he said. "You had some concerns about the new policy?"

"Yes, sir," I said, taking the chair opposite him. I noticed him getting out the same folder and touching the same black device as at our previous meeting.

"First of all, thank you for taking the time," I began. "I know this is a bit unusual."

"I'm happy to hear your concerns," he replied. There was a pause. I had mentally rehearsed what I was going to say on the drive over, but now that I was in the hot seat, my thoughts were in a jumble. Fragments floated through my mind: side effects, coercion, informed consent, abortion. I wasn't sure where to begin.

Come, Holy Spirit, I prayed. The chief was watching me intently, waiting. I took a breath and plunged in.

"Chief, I'm very concerned that the city is making this mandatory," I said, my words tumbling out fast. "It's not that I'm against the vaccine in principle, it's just that I have an issue with the brands that are available right now. I'm Catholic, as you know, and the fact that these shots were made using aborted fetal cells is a big problem for me. There's a good chance there'll be other brands coming out soon, ones that don't have a link to abortion. I would be fine with taking those, but right now I don't have any options that don't go against my beliefs."

It was not how I had planned to begin. I had imagined myself giving an articulate appeal for employees' rights, and working in my own beliefs later. In my nervousness, my primary concern had come spilling out. I had been looking down at the table while I spoke, and now I glanced up. What I saw made me falter. Forbes was fixing me with an unblinking stare. All warmth had left his features, and the look he directed at me was like ice. He said nothing, apparently waiting for me to continue.

"The . . . the other thing is," I plunged on, trying to gather my thoughts. "Has the city promised any kind of compensation if someone gets disabled because of the vaccine?"

He nodded briefly. "I believe they have."

"Because I know of a lot of people who have been injured and even killed by it. I know that's anecdotal, but I don't think we can ignore it, either. These aren't just stories I've read on the internet. These are real people in my circle of family and friends, people who were healthy until they took the vaccine. Chief, this thing is so new, and we don't know enough about it. I'm just not confident that we've made a proper . . ." I paused, searching for the right term. "A proper risk-benefit analysis," I finished. "I mean, can the city really make employees get an experimental medical product, and then absolve themselves of all responsibility?"

I paused again. He held me with the icy stare for another few seconds, then sucked in through his teeth.

"Well, Ben," he said. "I'm not a doctor, and I'm not a lawyer. I can't really answer those questions. I just know that city legal has advised us that we are within our rights as management to require this."

"But," I blurted out vehemently, "couldn't it be classified as assault to force someone to put something in their bodies that they don't want?" He sat up a little straighter.

"We're just talking here," he said edgily. "That's the kind of question you can bring to your own attorney."

"Chief, I just think I need more time," I said, in a slightly calmer tone. "I need time to do my research, see what these vaccines do to people in the long run, wait for ethical brands to become available."

"Well, we're almost two years into the pandemic," he replied. "You've had plenty of time to do your research.

And there are plenty of doctors out there who contradict your beliefs."

I felt myself redden at this condescension.

"I know several doctors who don't recommend it," I said. "Including my family doctor."

"Well, I'm not going to debate the vaccine with you," he said firmly. "All I can say, as a chief, is that we're going forward with this. It looks like you need to make a choice, Ben. My recommendation would be to talk to your union. Maybe they can help you."

He stood. "I'm sorry you had to come all the way to headquarters," he said, in a tone that may have been meant to convey understanding. "These times haven't been easy on anyone."

As I left the building, real anger surged in my gut for the first time.

"So that's how it is then," I muttered to myself. "All right, Chief, you've got yourself a fight."

When I got home, I could tell that Kate was upset.

"What happened?" I asked.

"I took the girls to swimming lessons," she said indignantly. "They wouldn't let us in the door. No one allowed without a vax passport, they said. Marie was in tears. It's so wrong! So wrong!"

I sighed deeply. "What's happened to our country?" I said.

"I've never been treated like a second-class citizen before," she said.

"Did they expect the kids to be vaccinated, too?" I asked.

"No, but I'm worried that we might be facing that in the near future. If they're forcing you to take it at work, how soon do you think it'll be before they come after the children?"

"We have to stop this before it gets that far," I said. "The question is, will enough people take a stand?"

~

I disconnected the battery of the dented minivan, while Clay attended to the occupants. They were standing on the sidewalk, smoking cigarettes, a bit shaken, but otherwise unharmed. The owner of the other vehicle was deep in conversation with a police officer, holding up his license and registration. Rescue 16 had also responded to the call, but with no extrication required, they were preparing to clear.

There was nothing to do now but wait for the tow truck.

"Hey, can I talk to you?" Bryce was standing in front of me.

"Sure," I said. We walked a few paces away from the accident scene and stood behind the tail board of the engine, sheltered from traffic and eavesdroppers.

"Did you come to a decision?" Bryce began.

"Yes," I replied. "I'm not going to get it."

"Are you going to try for an accommodation?" he asked. "The deadline's coming up soon."

"I'll try," I said. "I was planning on talking to Ted this week."

"Well, if anyone's eligible for a religious exemption," he said, "it's you. We all know how strong your beliefs are."

I nodded, unsure how to reply.

"What about you?" I asked.

A tired look crossed his face.

"You know what," he said quietly. "I went out . . . and I got it. I can't afford to lose my job."

There was a sadness in his voice.

"Everyone has to make the choice that's best for them," I said. It was all I could think to say, and it felt hollow.

He smiled mirthlessly.

"Thanks," he said. "Let me know how it goes."

Later that day, I made the call to Ted English. His voice was clipped and informative on the other end.

"I can tell you right now, we're not going to fight the mandate," he said.

"Why not?" I asked.

"Because we worked so hard to get firefighters to the front of the line for the vaccine," he replied. "If we come out against it now, we'll be a laughingstock."

"Other fire unions are pushing back," I said. "I can name several major cities that have gone up against management on this one."

"Yeah, but those other cities have hundreds of unvaccinated firefighters. There's only about twenty of you here, and your numbers are shrinking every day."

"Where does that leave us then, if you won't represent us?"

"I'm willing to represent you . . . individually," he said defensively. "I'll do everything I can to help you get a workplace accommodation, but I'm not going to fight the chief on the vaccine."

"So what are my options?" I asked.

"Well, you could always submit a request for exemption. It can be on health grounds, if you're worried about side effects, but I'll warn you that a health-based exemption isn't likely to go anywhere. We've already had about a dozen denials."

"What about religion?" I asked.

"If there's something in your religion that says you can't get a vaccine, it's worth a try," he offered.

"Well, I'll go for that," I said.

"You write a letter stating your creed-based objections," he said. "And I'll have the bargaining committee look it over. If it's something they think has a chance of success, we'll submit it to management on your behalf. Who knows, maybe we can get you an accommodation so you can remain in your position, a Covid test at the start of each shift or something. There's gotta be some way you can keep working."

I set out to compose the letter right away. I pored over the *Catechism of the Catholic Church*, papal encyclicals, and various other writings by Church leaders on abortion, vaccines, moral cooperation, and human freedom. By the end of the week, I felt I had built a strong case. I sent the draft off to Ted, and after two days I received the bargaining committee's list of recommendations. There were key words to insert, and certain phrases to be used, to convince the city lawyers that my objections were legitimate. I applied these edits dutifully, submitted the final copy, and waited. Ted called me later in the day and informed me that the union had agreed to send it up to city hall. A few days later, I got a letter in the mail. It was from the Human Resources department. I ripped it open eagerly. It bore the city letterhead, a case number, my name, and these few lines: "We have reviewed your exemption request and have determined that you do not meet the eligibility requirements. Exemptions are granted on the basis of legitimate need, not because an employee does not wish to get vaccinated. Please review the vaccination policy for more information." It was signed simply "HR Staff." I threw it down, alternating between disbelief and disgust. Had they even read my letter? How had a three-page defense been reduced to the innocuous "does not wish to get vaccinated"? And no one had been willing to put his signature to the denial.

Cowards, I thought. The deadline was coming up in a week. I had two shifts left. I called Ted.

"They denied it," I said irritably.

"I thought they might," he replied. "No one seems to be getting an exemption these days. We can always submit a grievance, but with the way things are, it's a toss-up whether an arbitrator will side with you. Basically, you'd have to prove your religion expressly forbids you to take a vaccine. And that's every vaccine, not just Covid."

"Well, it doesn't do that," I said.

"It would be a hard sell, then," he replied. "Your bishops have made allowances for getting an abortion-derived vaccine, which makes any kind of creed-based objection difficult."

"I know," I said. "But what about conscience rights?"

"Our lawyers have already advised us that conscience-based objections will go nowhere. It's too nebulous."

I sighed. "Then we're out of options," I said.

"At this point," he advised, "your best bet is to take the leave of absence they give you, and hope the political landscape changes before your employment review."

"Well I'm not going to let them fire me," I said. "I'll send in a letter of resignation first."

"Don't be too hasty about that. You'll have four months. And who knows, if case numbers drop they may end up scrapping the mandate altogether."

"We can only hope, I guess," I said. "In the meantime, I'm in a bad situation."

"I know," he replied. "But unless you change your mind and get vaccinated, it's your only option."

"I'm not going to do that," I said doggedly.

"Well, I'll certainly do whatever I can for you," he said. "I'd hate to see a brother lose his job."

The first person I saw when I walked into the station the following morning was Frank. I expected his usual hail-fellow-well-met greeting, but I was surprised when all he offered was a curt nod. A little later, the two of us were alone in the kitchen, and he began to speak.

"You still haven't got the shot, have you?" he asked bluntly.

"No," I said.

"I've been thinking about this," he said. "I'm not sure what this is going to mean, with you being the only one in the station that hasn't got it."

My heart sank. *Not you, Frank,* I thought.

"You mean you don't want to work with me?" I asked with some annoyance. He thought for a moment.

"I don't know," he said eventually. "Don't take this the wrong way, it's just, I've got my family to think of. If I get something from you at work, well, that would be bad."

"If these vaccines work as well as you say they do," I objected, "why are you worried about catching something from me?"

"I've been doing some research," he replied, "and unvaccinated people spread it more easily."

"I've been doing some research, too!" I shot out, my temper rising. "And it's illegal to discriminate against people for their medical status! You wouldn't corner someone and ask them 'Do you have AIDS?' and then refuse to work with them if they said yes!"

Frank looked very taken aback. In the two years we had worked together, I had never once raised my voice at him.

"Sorry," I said.

"It's okay," he replied, looking at me keenly. "I hear ya. I guess there's privacy laws and stuff like that. I was just thinking out loud. It's okay, don't worry about it."

"If it would make you feel better, I can wear a mask," I said.

"No, no," he said. "You don't need to do that. Forget I said anything." He laughed. "What a conversation first thing in the morning!"

"Frank, you know I only have two more shifts, and then I'm out of your hair," I said quietly.

His face fell.

"Ugh, I hope it doesn't have to come to that," he said.

"Unfortunately, it's looking that way," I replied.

"I feel bad for you, buddy, but honestly, I can't say I understand your decision. I think everybody should get vaccinated. That's just my opinion."

"Should I get fired for my opinion?" I asked.

"No-o," he said slowly. "And I get there's a religious aspect to it, but still . . . "

"Honestly, it's about abortion. They made these vaccines with the remains of little babies. I just can't bring myself to support that."

"Well, the thing about abortion is," he said stolidly, "if a woman gets raped, she shouldn't be forced to have a baby."

"Rape's a horrible crime," I agreed. "And rapists should be thrown in jail. Here's my question: Why do we still give the death sentence to the innocent child?"

He cocked his head, shook it, then said, "You and I could argue about this all day long."

"That's the thing," I said. "We can disagree, and none of us will get fired for our views. But when it comes to the vaccine, it looks like only people with one opinion get to keep their jobs."

"I dunno, buddy," he said. He stood and headed for his office. "What a conversation!" he exclaimed as he walked away.

Clay and Tyler took it in stride when I told them that the next shift would be my last.

"I wouldn't assume that," said Tyler optimistically. "They will probably let you come in Friday morning, same as usual. This whole thing was just about getting as many people to get the shot as possible. They wouldn't really kick anyone out."

"Would be nice," I said. "But from what I'm seeing, they mean business. I'll bring in a cake next shift."

"Make it a good one," said Clay.

The letter was sitting in my mailbox when I got home the following morning. I opened it with Kate in the living room. It bore the same city letterhead, and was signed by Chief Forbes:

> According to our records, you have not submitted proof of vaccination in accordance with the city's mandatory vaccination policy. Further to the requirements of the policy, you will be placed on an unpaid leave of absence, to commence from the next Friday that C Shift is scheduled to work. The leave will be effective for a period of four (4) calendar months from that date. If we still have not received your proof of vaccination by the time your leave has elapsed, we will conduct a review of your case, which may result in termination of your employment.

We looked at each other.

"Well," I said at length. "We chose this."

"We didn't choose it, they brought it to us," said Kate.

"What happens now?" I asked.

"God will take care of us," she said.

"We've been talking about taking a leave of absence for your health. Maybe this is an answer to prayer."

"I know, but we can't travel right now, not without vax

passports. Maybe we need to think about using this time to fix up the house."

"Selling, you mean?"

"I don't know if that's what we're supposed to do, but if you lose your job, I don't see how we can afford to stay here."

"Let's not give up hope yet. If I push them hard enough, maybe there's a chance they'll relent and let me come back to work."

"Try, but let's be open to whatever God wants to do with us."

Forty of us stood in the chill November air, waiting for the doors of the police station to open. An icy drizzle speckled my shoulders, and I tucked the cardboard folder I was carrying carefully under my arm to keep it dry. Inside was a sheaf of papers, with the names of various city managers, from the mayor down to the fire chief, and a list of charges to be brought against each of them: violation of privacy, breach of civil rights, denial of informed consent. It was a last-ditch effort, but there was a feeling of determination about the small crowd of city workers who had gathered that morning. We were a motley collection of cast-offs, men and women of all ages, from various city departments, all out of work and all united in resistance to the mandate.

My mind went back to the morning before. I had shaken hands with Frank and Clay, and Tyler had gone for the bear hug.

"We'll see you in a few months," he said cheerfully. "I know you'll be back."

"There's always a spot for you at Station 8," agreed Frank.

"Be good," said Clay. "Call me if you need anything."

Then I had walked out of the station for the last time.

"Drop in for lunch some time," Tyler called after me.

"I wish I could," I replied. "I've been told I'll get a trespassing charge if I'm caught on city property."

Now I stood in the rain, clutching my bundle of papers, hoping for a last chance of saving my job. As I looked around me, I noticed two young men standing near the door. They were well-dressed, with military haircuts. I walked over and introduced myself.

"What department do you guys work for?" I asked.

"We're paramedics," answered the shorter of the two. He wore a grave expression. "I'm Conor, this is Michael. Thanks for joining us."

"No problem. Nice to meet you both," I said. "Do you think we have much longer to wait?"

"I hope not," Conor replied. "I was in there about half an hour ago. They *are* open, so there's no reason they shouldn't let us in. The constable I talked to told us to wait outside. They're not sure how to handle so many people."

"That doesn't bode well," said his partner.

"They can't *not* let us press charges," said Conor. "That would be obstruction of justice. Hopefully we get an answer soon."

We stood around for a few more minutes, and the conversation lagged.

"I've been hearing a lot about vaccine injuries," I said at length. "But I haven't seen anything firsthand. Have either of you come across it?"

Conor smiled grimly. "Man, before I got put on leave, I was transporting at least one patient per shift with a vaccine injury, sometimes more."

"You're kidding!" I exclaimed. "So it's not just conspiracy theory?"

"No," he said emphatically. "And we're not talking sore arms, either. I mean strokes, paralysis, heart problems, big stuff. All within days of getting the shot. Of course when they tell my other colleagues that, they brush it off. I heard my supervisor tell a patient his chest pains were happening because he was just 'nervous' about the vax!"

I shook my head.

"So you organized this?" I asked.

"Yes," he replied. "I'm hoping we can all work together. The various departments don't always communicate with each other very well, but if we can stick together on this one, maybe we can get something done."

Other people were gathering around us now, impatient to get out of the rain.

"I'm sorry, everyone," said Conor, raising his voice. "I don't know how much longer we have to wait. Hopefully they will let us in soon."

I mingled with the crowd, meeting a few people and listening as some shared their stories. One older man had driven a garbage truck for twenty years. He was now seeing his pension and hopes of retirement disappearing.

"I had a big scare after my first shot," he confided. "I was in a lot of pain, and I started swelling up all over. I swore I'd never get a second. Now I'm out on my ear."

One young woman, visibly pregnant, introduced herself as an ultrasound technician.

"Of course I can't get the shot," she said. "I told my supervisor I couldn't, because of the baby, but it didn't matter. Luckily my husband has his own company. I don't know how we'd get by otherwise."

There were at least a dozen nurses of various ages. One

was a motherly type who must have been close to retirement.

"They've laid off almost three hundred people between the five hospitals," she sighed. "And then they're complaining about our health care system being overloaded. If you ask me, they've lost their minds."

"Three hundred! I wish they'd all showed up today," I observed.

"I don't think many people are willing to go out on a limb like this," she replied. "Pressing charges against your boss isn't something one does every day."

I shook hands with four strong-looking men with serious faces. In the course of the conversation, it turned out that they were firefighters from other shifts, none of whom I had met.

"Is this everyone from Fire?" I asked, with some disappointment.

"I don't think there were that many of us to begin with," one of them answered. "By the way, I am vaccinated," he added, almost apologetically. "I'm here because I take issue with the city making me disclose it. It's a breach of privacy."

"So you're still working then?" I asked.

"Yes," he replied. "I'm the captain at Station 1 on B Shift. I've had lots of arguments with the chief," he chuckled. "He doesn't like me much."

"I don't think he likes me, either," I said. "Not after I told him the mandate amounted to assault."

He laughed.

"Yeah, that wouldn't have gone over well. You know his mom died of Covid last year?"

"I didn't know that," I said. Suddenly, my conversation with Forbes was making more sense.

"Yes, he blames the unvaccinated for it," the captain went

on. "There's nothing you can say that will make him hear you."

At that moment, the door opened, and a police officer stepped out. He had a shaved head and a sleeve of tattoos up one arm.

"All right!" he shouted. "Thank you all for your patience! I've been consulting with my sergeant about you guys, and he wants me to tell you that we won't be able to process your complaints today!"

A murmur rippled through the crowd. People pressed forward, many with hostile expressions. One of the nurses pulled out a cell phone and began filming.

"What do you mean, you can't process us?" demanded Conor. "This is a normal procedure, isn't it? People come and file charges all the time!"

The officer shook his head emphatically.

"What you're bringing aren't criminal charges," he said. "It's more of a civil matter. I suggest you take up your complaints with your own managers."

"We've done everything we can with our managers," said Conor, forcing his voice to stay calm. "They are doing something illegal, and now it's time for the police to do their due diligence and look into it."

"I'm sorry," the officer replied. "My sergeant is adamant that it's none of our business."

"Then can you send your sergeant out here?" snapped Conor. "I'd like to speak to him myself."

"I could ask," he said skeptically. "Hold on." He disappeared inside.

"I was afraid of this," Michael the paramedic commented grimly.

"Ridiculous!" fumed one of the nurses.

The rain was coming down harder now. In a minute, the

door opened and the officer reappeared, accompanied by a tall, silver-haired man in uniform.

"Hi, everyone," the sergeant began in measured tones. "So I understand you're here today with an issue to do with the vaccine mandate, is that correct?"

"Yes," said Conor.

"Unfortunately, the police do not handle complaints of that nature," he said. "This is a purely civil matter."

One of the firefighters spoke up.

"Are you familiar with the definition of extortion?" he asked.

"Um, maybe not the exact definition," the sergeant replied. "It's not a charge we hear very often."

"Well, let me read it to you," the firefighter said, pulling out a sheet of paper.

"We don't have time for that," the constable with the shaved head said impatiently.

"Let him read it," said Conor.

"Fine," the constable barked. "Read it!"

"Extortion is the use of force, threats or intimidation to extract money or property from someone else," the firefighter read. He looked up at the officers. "Don't you see? That's what's going on here. Our managers are trying to intimidate us into revealing personal and private medical information—that's our intellectual property. And they're using this information to strip us of our livelihood. They're forcing us to take a harmful product and threatening us with our jobs if we don't comply. That's extortion!"

"That's a stretch," the constable snorted.

"We're just trying to save our jobs," said Conor. "All these people are wondering whether they'll be able to feed their families a month from now."

"Have you seen what's happening in other countries?"

came a woman's voice from the back of the crowd. "They're fining and arresting unvaccinated people. We're trying to stop that from happening here. Are you okay with those kinds of measures?"

"Well," the constable huffed, "we could play 'what if' indefinitely here. I don't see what bearing that has on—"

"What would be the line for you?" Michael interrupted in a loud voice.

"What do you mean?" he retorted.

"Up to what point would you enforce laws that violate civil rights? If you were ordered to go door to door arresting unvaccinated people, would you do it?"

He looked exasperated. "Listen, I'm not here to discuss politics. I understand what you guys are trying to do, and I sympathize, but we just can't accommodate that right now. I'm sorry!"

"It's time to break up this gathering," the sergeant said. "Please leave the premises."

They turned and walked back into the building.

~

On an overcast March afternoon, I closed the farmhouse door and turned the key for the last time. The minivan and the pickup stood waiting on the driveway, loaded with the few things we had decided to keep. I took one last look around before climbing into the cab of the truck, where the girls were already buckled in. Kate sat behind the wheel of the van, ready for the little procession to depart.

"All set, girls?" I asked.

"Yes!" they replied enthusiastically.

As we drove slowly up the long driveway, they waved at all the familiar objects.

"Good-bye, house! Good-bye, swings! Good-bye, barn! Good-bye, climbing trees!" they sang in unison.

I felt a lump in my throat. If things had been different, this would have been our homestead. Here we had planned to weather the years, see our children grow up, and grow old ourselves. We had put down roots here. Our dreams for the future had joined the many sacrifices of the past four months. And yet, despite the tug of sadness at my heart, I felt lighter than I ever had before. It was a strange paradox. Here I was leaving behind everything I had believed would bring me happiness: the career of my choice, the farm I had always wanted, the possessions I had built up around me to create comfort and security. They were gone, and I was much the better for it. I drove past the "Sold" sign at the end of the driveway and turned onto the main road.

"Are you looking forward to living at Grandma and Grandpa's?" I asked.

"Oh, yes!" they chorused.

"Are you sad about leaving our house?"

"Yes," Kathleen replied. The other two were silent.

"That's a normal thing to feel," I said. "I feel sad, too. But we have a great adventure ahead of us, don't we?"

"Yeah," they replied, with nods.

"Are we always going to live with Grandma and Grandpa?" asked Beth.

"Just until we can find a new house," I replied. "Once Daddy gets another job."

As the miles rolled by, the sadness lifted, and I felt a growing anticipation for the next stage of my life. There were many uncertainties ahead, but one thing I had learned through the letting go was that God is always faithful. Faith was a little bit like a parachute: you had to jump out of the plane before you knew whether it would open or not.

We had just jumped, and it remained to be seen what God would do with our choice.

A few weeks later, an email came from Ted English. I had been checking messages on my phone, and now I sat down heavily in the nearest chair, scarcely able to believe what I was reading.

> Great news Ben!
>
> Just wanted you to know that I got off the phone with John Forbes. The city has rescinded its vaccination policy. Next week, unvaccinated members will be permitted in the workplace. A couple of weeks ago we met with the city legal team and wanted answers on why other cities were using testing as opposed to vaccines. They stated that they would find that data and get back to us . . . and now they have rescinded their policy.
>
> I am truly sorry that yourself, your family, and others had to deal with this nonsense and never got the answers you were deserving of by the city. I hope that you and your family are doing well given everything that has happened. I will say that I am truly lucky to have been dealing with you, seeing how strong your convictions and beliefs are. That is something to really be proud of! I would have kept fighting for you for every extra second out of this city. If you still choose to get vaccinated that is of your own free will at this point. I am always available to talk anytime, to chat or listen.
>
> Sincerely,
> Ted

Sunlight filtered through the station windows. Tyler was serving up a batch of eggs, while Frank and Clay sat at the kitchen table, sipping their morning coffee. I filled my mug and pulled up a chair, grinning a little.

"Glad to be back?" asked Frank.

"Yes," I replied. "I am."

"I always knew you'd be back," said Tyler, placing a steaming plate in front of me. "Here, eat up."

He put two more down in front of Frank and Clay. Frank took a giant mouthful.

"Dang, that's good!" he exclaimed.

"It's all right," said Clay with a twinkle in his eye. I laughed. This was like old times.

"There's the printer!" said Tyler eagerly. We were up and running for the truck.

As Engine 8 left the suburbs behind and raced down country roads, I said a little prayer.

Lord, I haven't done this in four months. Help me to remember my training and do a good job.

We were getting our first updates.

"Engine 8, from Dispatch. Be advised, this is for a dump truck versus an SUV. Bystanders are saying we have an entrapment."

"We'll be the first ones there," said Frank. He picked up his radio mic.

"Dispatch, from Engine 8. Copy that. Please have Rescue 16 respond on this one as well."

We came around a corner and saw the accident scene in front of us. A dump truck had pulled off to the side of the road, its front bumper dented in and a streak of leaked coolant leaving a dark trail across the pavement. On the other side of the road, straddling the ditch, was the SUV. Its front end was completely destroyed, and mangled car parts littered the roadway. The windshield was a spider web of cracked glass, and I could see a human figure slumped over the steering wheel.

"I'll take patient care," I said, grabbing the med bag.

"I'll stabilize," said Tyler, reaching for the tool pouch.

The moment the truck came to a stop, I was out the door and hurrying toward the car. As I approached, I noted the position the vehicle had landed in. Judging by the damage, it had struck the dump truck head on, and the force of the impact had propelled it into the ditch. It now sat at right angles to the road, with its front wheels resting on the shoulder, and the back wheels on the lawn of a neighboring house. The ditch yawned under it, a good five feet deep. I approached the driver's window. The occupant was an older woman, probably sixty years of age. I could see that she was breathing slightly, and there was a large pool of blood soaking her left leg. The window was down, and I reached in and tapped her on the shoulder.

"Hey!" I called. "Can you hear me?"

There was no answer. I pinched her shoulder as hard as I could and was answered by a hollow groan. My fingers went to her carotid artery. There was a pulse, but it was thready and weak. I performed a quick head-to-toe exam, feeling for cuts or broken bones. When I got to her leg, she shuddered and emitted another groan, louder than the first. Something didn't feel right about that leg, either. A section of the upper thigh moved under my fingers.

"What have we got?" Frank was standing on the shoulder, assessing the scene.

"Responds to painful stimuli," I replied. "Weak pulse. There's a bad cut on her leg. Might be broken."

"All right," he said. "Let me know right away if her condition changes. We've got to stabilize this vehicle before we can get her out."

Tyler had already chocked the tires and was now working away with a chain, tying off the vehicle to a nearby tree. Clay was busy ferrying up the extrication tools. There was a crunching of tires on gravel as an ambulance pulled up. I rummaged in the med bag and found three bulky dressings. I removed them from their sterile packaging and reached in the window. The patient's eyes half-opened for a second, and she muttered something inarticulate.

"Here!" I said loudly. "Hold these for me!"

Her eyes flickered wider for an instant, then closed.

"Hey!" I shouted. I seized her left hand and placed a dressing in it, then clamped it over the wound.

"Hold that for me!" I instructed. "We're going to get you out of here in just a few minutes."

She may have been able to understand a little, for her hand stayed in place. A siren sounded, growing louder. Rescue 16 was approaching. Footsteps sounded on the shoulder, and a paramedic came up to the window beside me.

"What have we got?" he asked.

"Elderly female, in and out of consciousness," I replied. "Bad cut on her leg, possibly broken. She's breathing okay."

He reached in. A shriek erupted from the vehicle.

"It's broken all right," he said. "The bone's poked through the skin then gone back in. We gotta get her out of here ASAP."

"Right," I said. "Hey, Captain!" I waved Frank over. He came hurrying. I could see Rescue 16 pulling up behind him.

"She's got an open fracture," I informed him.

"Right," he nodded. He pulled out his portable radio.

"Dispatch, from Engine 8. We have an unconscious patient with an open fracture to the leg. We will begin extrication operations immediately. You can show Rescue 16 on location at this time. Patient may not have long, as there is the possibility of massive internal bleeding. Please send an air ambulance."

I turned to the paramedic.

"You'll take over patient care?" I asked.

"Yes," he replied.

I moved with urgency over to the extrication tools. I set the choke on the portable motor, pulled the ripcord, and heard it splutter into life. I hooked the hoses up to the spreaders, while Tyler did the same with the cutters. The crew from Rescue 16 were walking toward us, carrying long telescoping jacks to stabilize the vehicle further. I recognized Gino in the lead, and two overtime firefighters who were strangers to me. They set to work, propping the vehicle up from underneath. Tyler and I put down our tools beside the driver's door and began removing glass and peeling away trim to reveal any wires or airbag cylinders. That done, I lifted the spreaders, waited for the nod from Frank, and began taking the door. The metal groaned a little as it peeled away from the latch, but everything was going smoothly at the start. There was a silent prayer on my lips.

Help me to get this door off, Lord.

When the metal began to tear, I stopped. I closed the jaws a little and dropped them down farther into the gap. One quick spread, and the door popped off. Tyler quickly cut the hinges, and two rescue firefighters dragged the severed door off to the side. Next we tackled the rear door, and, after some difficulty, it, too, was carried away. Another ambulance had arrived, and a crew of paramedics was tending to the patient, preparing her for removal from the car. Tyler took over the cutters, and in minutes he had removed the thick B-pillar to which the doors had been fastened. The side of the car was now completely open, and we had a large space to work with.

"I think we're ready to pull her out," one of the paramedics said. "Here," he directed me. "You take her legs. We're going to turn her and lower her onto the backboard."

I hadn't been looking at the patient for the last few

minutes, and when I crouched down beside her, what I saw shocked me. Her leg had shifted during the course of the operations, and a long shaft of bone protruded from her thigh. It was sticking out a good four inches, and the end had snapped off cleanly. The leg muscle was contracted tightly around the bone, and no blood came from the wound, though clearly it had been bleeding earlier, as evidenced by her soaked pant leg.

The paramedics slid a backboard under her, and together we worked her slowly around. She groaned and mumbled a little as we lowered her onto the board, but otherwise she gave no sign that she was aware of what was going on. There was a humming sound in the distance. As we worked, it grew louder, and soon the air was filled with the throbbing of an aircraft engine.

"Here comes the chopper!" the lead paramedic announced. "All right, let's get her up to the road!"

Paramedics and firefighters surged around the patient, hands reached out to grasp the board, and with a sharp command we were lifting her and carrying her up the bank. The noise was deafening now. A helicopter appeared above the treetops, advanced slowly until it was hovering over the road, and began a gradual descent. A blast of wind from the rotors hit us. Leaves and sand flew in all directions. I blinked and turned my face away. After a few seconds, it subsided, and the air ambulance settled gracefully onto the road. We reached the top of the bank just as it was shutting off its engines.

"We'll take it from here," the lead paramedic said. "Thank you for all your hard work."

We stood in a group and watched them wheel the patient toward the waiting helicopter. A strange feeling stole over me. It was like warmth, and with it came a sense of accomplishment and peace that I had never known before.

I'm back at work, I thought to myself. *After everything, I'm back*.

As we returned to the station, Clay belted out a country song off-key.

I laughed aloud. God had been faithful, and he had led me through adversity, back to where I belonged. And yet, it was not the job itself that brought the happiness. I realized that true peace came from being in God's will, wherever it might lead. My thoughts strayed back to my brother's poem:

> I shall be telling this with a sigh
> Somewhere ages and ages hence:
> Two roads diverged in a wood, and I—
> I took the one less traveled by,
> And that has made all the difference.

ACKNOWLEDGMENTS

I would like to thank the following people for their valuable editorial suggestions: my father, Michael D. O'Brien, my father-in-law, Thomas Anderson, and my wife, Katelyn O'Brien, without whose encouragement and support this book would not have been written.